FROZEN CHOSIN

U.S. Marines at the Changjin Reservoir

by Brigadier General Edwin H. Simmons, USMC (Ret)

The race to the Yalu was on. General of the Army Douglas MacArthur's strategic triumph at Inchon and the subsequent breakout of the U.S. Eighth Army from the Pusan Perimeter and the recapture of Seoul had changed the direction of the war. Only the finishing touches needed to be done to complete the destruction of the North Korean People's Army. Moving up the east coast was the independent X Corps, commanded by Major General Edward M. Almond, USA. The 1st Marine Division, under Major General Oliver P. Smith, was part of X Corps and had been so since the 15 September 1950 landing at Inchon.

After Seoul the 1st Marine Division had reloaded into its amphibious ships and had swung around the Korean peninsula to land at Wonsan on the east coast. The landing on 26 October 1950 met no opposition; the port had been taken from the land side by the resurgent South Korean army. The date was General Smith's 57th birthday, but he let it pass unnoticed. Two days later he ordered Colonel Homer L. Litzenberg, Jr., 47, to move his 7th Marine Regimental Combat Team north from Wonsan to Hamhung. Smith was then to prepare for an advance to the Manchurian border, 135 miles distant. And so began one of the Marine Corps' greatest battles—or, as the Corps would call it, the "Chosin Reservoir Campaign." The Marines called it the "Chosin" Reservoir because that is what their Japanese-based maps called it. The South Koreans, nationalistic sensibilities disturbed, preferred—and, indeed, would come to insist—that it be called the "Changjin" Reservoir.

On the Cover: *In this poignant photograph by the peerless Marine and* Life *photographer, David Douglas Duncan, the dead ride in trucks, legs bound together with pack straps.* Photo by David Douglas Duncan

At left: *A cold and exhausted Marine light machine gunner climbs resolutely toward a ridgeline.* National Archives Photo (USAF) 342-FH-37885

General Smith, commander of the Marines—a quiet man and inveterate pipe-smoker (his favorite brand of tobacco was Sir Walter Raleigh)—was not the sort of personality to attract a nickname. His contemporaries sometimes referred to him as "the Professor" but, for the most part, to distinguish him from two more senior and better known General Smiths in the World War II Marine Corps—Holland M. "Howlin' Mad" Smith of famous temper and mild-mannered Julian C. Smith of Tarawa—he was known by his initials "O. P."

Across the Taebaek (Nangnim) Mountains, the Eighth Army, under Lieutenant General Walton H. Walker, was advancing up the west coast of the Korean peninsula. Walker, a short, stubby man, was "Johnnie" to his friends, "Bulldog" to the press. In World War II he had commanded XX Corps in General George S. Patton's Third Army and had been a Patton favorite. But these credentials held little weight with General Douglas MacArthur. He had come close to relieving Walker in August during the worst of the situation in the Pusan Perimeter. Relations between Almond and Walker were cool at best.

Department of Defense Photo (USMC) A88898

MajGen Oliver P. "O.P." Smith commanded the 1st Marine Division throughout the Chosin Reservoir campaign. A studious man, his quiet demeanor belied his extensive combat experience. His seemingly cautious style of leadership brought him into frequent conflict with MajGen Edward M. Almond, USA, the impetuous commanding general of X Corps.

MacArthur had given Almond command of X Corps for the Inchon landing while he continued, at least in name, as MacArthur's chief of staff at Far East Command. Almond, an ener-

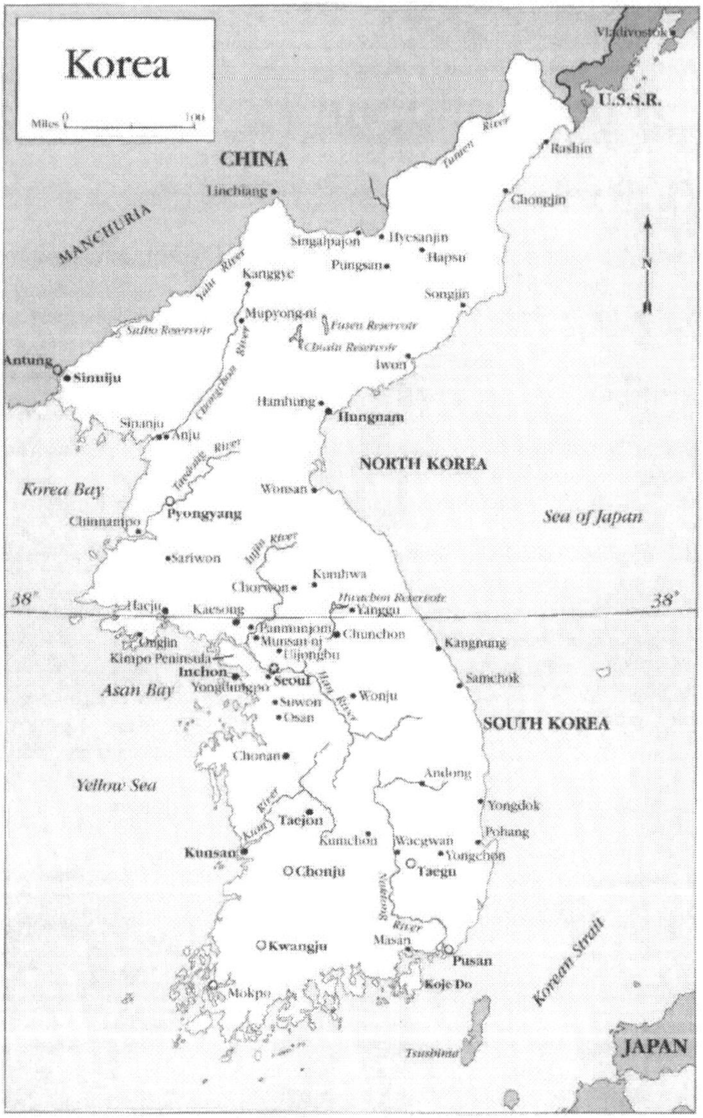

getic, ambitious, and abrasive man, still nominally wore both hats although his X Corps command post in Korea was a long distance from MacArthur's headquarters in Tokyo.

"General Almond in 1950 and 1951 in Korea had several nicknames," wrote Roy E. Appleman in his *Escaping the Trap: The U.S. Army X Corps in Northeast Korea*. "Generally, he was known to his friends and close associates as Ned. Other names were 'Ned, the Anointed,' which meant he was a favorite of General MacArthur's, and 'Ned, the Dread,' which referred to his power, his brusque manner, and sometimes arbitrary actions."

Many persons, both then and later, thought that X Corps should have now been subordinated to the Eighth Army. But on the 28th of October, "O. P." Smith was less concerned with these higher-level command considerations than he was with events closer to his headquarters at Wonsan. The 1st Battalion, 1st Marines, sent south of Wonsan to the coastal town of Kojo-ri, at the direction of X Corps, to protect a Republic of Korea supply dump had been roughly handled by a surprisingly strong North Korean attack. Smith thought that the battalion commander "was in a funk and it would be wise for Puller to go down and take charge." Colonel Lewis B. "Chesty" Puller, the regimental commander of the 1st Marines, left late that afternoon by rail with the 2d Battalion, 1st Marines, "to clear up the situation."

On the following day, Smith was annoyed by an order from Almond that removed the 1st Korean Marine Corps (KMC) Regiment from his operational control. The two commanders conferred on Monday, 30 October. Almond agreed to the return of one KMC battalion in order to expedite the move of the 5th and 7th Marines to Hamhung. After meeting with Almond, Smith flew by helicopter down to Kojo-ri and found that Puller indeed had the situation well in hand.

Tensions and differences between Almond and O. P. Smith were no secret. Almond had first met the Marine commander on Smith's arrival in Japan on 22 August 1950. As Almond still asserted a quarter-century later: "I got the impression initially (and it was fortified constantly later) that General Smith always had excuses for not performing at the required time the tasks he was requested to do."

With the 1st Marine Division

Department of Defense Photo (USN) 421392

Six weeks after the successful assault of Inchon, the 1st Marine Division made a delayed but unopposed landing at Wonsan on 26 October 1950. Heavy mining of the sea approaches with Soviet-made mines caused the delay.

After landing at Wonsan, the 1st Battalion, 1st Marines, was sent south to the coastal town of Kojo-ri. Here it was savaged by an unexpectedly strong North Korean attack. MajGen Smith sent Col Lewis B. "Chesty" Puller south with the 2d Battalion, 1st Marines, to "take charge."

Photo by Cpl W T. Wolfe, Department of Defense Photo (USMC) A4323

assigned as part of X Corps, Almond was Smith's operational commander. Smith's administrative commander continued to be Lieutenant General Lemuel C. Shepherd, Jr., commanding general of Fleet Marine Force, Pacific, with headquarters in Hawaii. While not in the operational chain-of-command, Shepherd was responsible for the personnel and logistical support of the 1st Marine Division. This gave Shepherd, an old war-horse, an excuse for frequent visits to the battlefield. Fifty-four-year-old Shepherd and 57-year-old Almond got along well, perhaps because they were both Virginians and both graduates of close-knit Virginia Military Institute—Almond, Class of 1915 and Shepherd, Class of 1917. "I liked him," said Shepherd of

LtGen Lemuel C. Shepherd, Jr., left, arrives at Wonsan Airfield on 31 October for one of his frequent visits and is greeted by MajGen Edward M. Almond. The two generals got along very well, perhaps because they were both Virginia Military Institute graduates. Both are wearing cuffed Army combat boots, a footgear favored by all those who could obtain them.

Photo by Cpl Jack Nash, National Archives Photo (USA) 111-SC351739

Photo by Cpl Jack Nash, National Archives Photo (USA) 111-SC351740

Using a map spread out on the hood of a jeep, MajGen Edward M. Almond briefs LtGen Lemuel C. Shepherd on his arrival at Wonsan on 31 October. At the extreme left is MajGen William J. Wallace, Director of Aviation at Headquarters U. S. Marine Corps, who accompanied Shepherd. MajGen Oliver P. Smith stands behind Shepherd and Almond. MajGen Field Harris is on the extreme right.

Almond at VMI, "but I really can't say he was one of my closer friends." Shepherd had had reason to expect command of X Corps for the Inchon landing, but MacArthur had given command to his chief of staff, Almond. Shepherd exhibited no visible grudge and Almond, in turn, always made Shepherd welcome on his visits and he often stayed with Almond in his mess. Years later, Shepherd, who considered Almond an excellent corps commander, said of him: "He was energetic, forceful, brave, and in many ways did a good job under most difficult conditions."

Concerning Almond's relations with Smith, Shepherd said: "He and O. P. just didn't get along, from the very first. They're two entirely different personalities. . . . O. P. [was] a cautious individual, a fine staff officer who considered every contingency before taking action. On the other hand Almond was aggressive and anxious for the X Corps to push ahead faster than Smith thought his division should. Smith wisely took every precaution to protect his flanks during his division's advance into North Korea, which slowed him down considerably. I'm sure Almond got into Smith's hair—just like I'm sure that I did too."

As a glance at a map will confirm, North Korea is shaped like a funnel, with a narrow neck—roughly a line from Wonsan west to Pyongyang—and a very wide mouth, the boundary with Red China and a bit with the Soviet Union on the north, formed by the Yalu and Tumen Rivers. Because of this geographic conformation, any force moving from the north to south had the advantage of a converging action. Conversely, forces moving from south to north must diverge. As Walker and Almond advanced to the north, the gap between Eighth Army and X Corps would grow wider and wider. This may have concerned Walker, but it does not seem to have bothered Almond—nor their common commander, General Douglas MacArthur, many miles away in Tokyo in what he liked to call his "GHQ."

General Shepherd arrived at Wonsan for one of his periodic visits on Tuesday, 31 October. Next morning Shepherd flew down to Kojo-ri to visit Puller and on his return to Wonsan he and Smith flew to Hamhung to see Litzenberg. That night Smith entered in his log:

> Litzenberg is concerned over the situation. He has moved up behind the 26th

Col Homer L. Litzenberg, Jr., commanding officer of the 7th Marines, was known to his troops as "Litz the Blitz," more for the alliteration than his command style. At the outbreak of the war, Litzenberg was in command of the 6th Marines at Camp Lejeune. In August 1950, the 7th Marines was hurriedly re-activated at Camp Pendleton using cadres drawn from the 6th Marines.

National Archives Photo (USMC) 127-N-A4719

> ROK Regiment and will relieve them tomorrow. Two Chinese regiments have been identified to the front. The ROK regiment is very glad to be relieved by the Marines. The ROKs apparently have no stomach for fighting Chinese.

China Enters the War

Most Chinese historians now assert, and most Western historians are now ready to believe, that China entered the Korean War reluctantly.

In 1948, Kim Il Sung, then 37, had emerged under the patronage of the Soviet occupation as the leader of the so-called Democratic People's Republic of Korea with its capital in Pyongyang. In the West he remained a shadowy figure. Reputedly he had been a successful guerrilla fighter against the Japanese. He had returned to North Korea at World War II's end as a hero.

With the civil war against the Chinese Nationalists at a successful close, Mao in late 1949 and early 1950 released four divisions made up of soldiers of Korean origin to return to Korea. These Kim, under Soviet tutelage, reorganized into mirror images of Soviet rifle divisions in equipment and training. Early on Kim Il Sung learned how to play Mao Tse-tung against Stalin. For more than a year, Kim Il Sung zigzagged back and forth between Moscow and Peiping (not yet known in the West as "Beijing") seeking Stalin's and then Mao's support for an overt invasion of the South.

Both Stalin and Mao were at first skeptical of Kim's ambitions. Stalin cautioned Kim that he should cross the 38th Parallel only in a counteroffensive to a South Korea invasion of the north. Mao advised Kim to be prepared for protracted guerrilla warfare and not to attempt to reunify Korea by force.

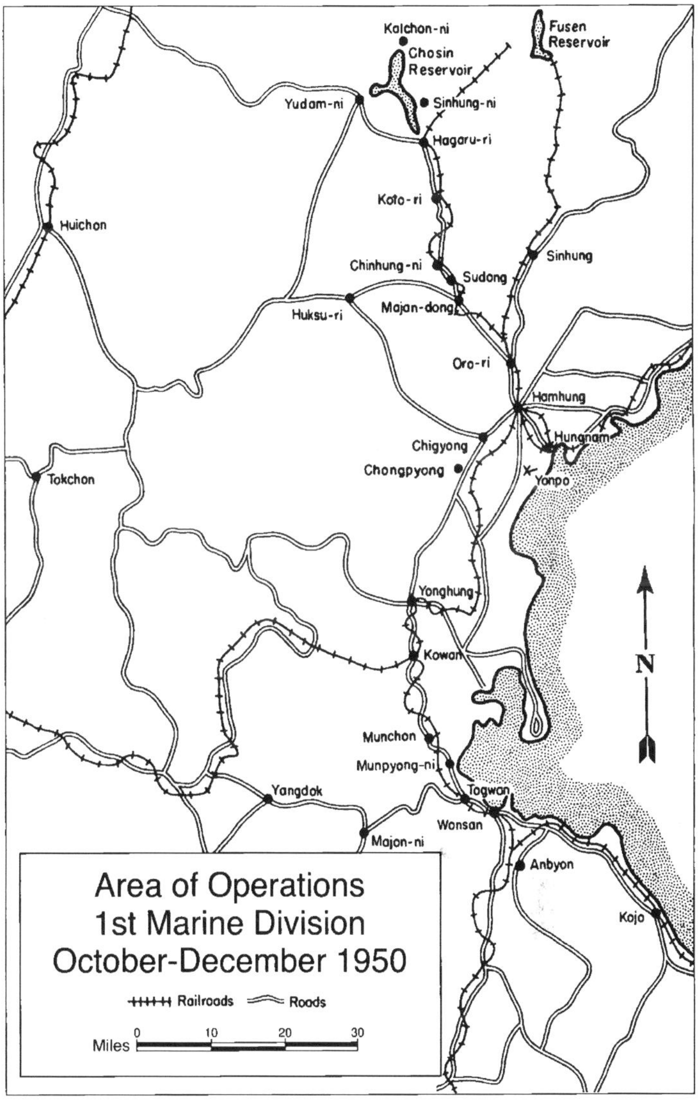

Area of Operations
1st Marine Division
October-December 1950

For Mao, Kim's ambitious plans were a distraction. He was much more interested in completing his victory against the Chinese Nationalists by "liberating" Xingjiang, Tibet, and, most importantly, Taiwan. But in the spring of 1950 Stalin, playing his own game, gave Kim a qualified promise of Soviet support with the proviso that the North Korean leader consult with Mao. Accordingly, Kim went again to Peiping in mid-May 1950, put on a bold front and told Mao that Stalin had agreed with his plan to invade South Korea. A cautious Mao asked the Soviet ambassador to confirm Kim's assertion. A sly Stalin replied that while he approved Kim's plans, the deci-

Marine Corps Historical Center Photo Collection

Peng Dehuai, left, commander of the Chinese Communist Forces meets with Kim Il Sung, premier of the Democratic People's Republic of Korea. Peng later was China's Minister of National Defense but because of his criticism of the Great Leap Forward he was dismissed. During the Cultural Revolution he was arrested, imprisoned, tortured, and died in 1974 from a lack of medical care. He was posthumously rehabilitated as a "great revolutionary fighter" and loyal party member.

sion to invade South Korea was a decision to be made by China and North Korea. Until then Mao's position had been that Taiwan's liberation must have priority over Korea's unification. Reluctantly, Mao reversed these priorities but cautioned Kim on the strong possibility of intervention by the United States on the side of South Korea. Mao had almost no regular troops in northeast China. He told Kim that he would move troops into Manchuria, poised to cross into Korea, but that they would not enter the war unless American troops crossed the 38th Parallel.

The concentration of Chinese forces along the Yalu did not really begin until mid-July 1950 with the formation of the *Northeast China Border Defense Force*, about 260,000 troops. By mid-August Mao was certain that the United Nations forces would land at Inchon. On 23 August, the same day that MacArthur was wresting final approval for Inchon at a conference in Tokyo, Mao was meeting with his political and military leaders in Peiping. They were ordered to complete all preparations by the *Northeast China Border Defense Force* for war. The *Ninth Army Group*, which had been poised near Shanghai for the invasion of Taiwan (still known to the Western world as "Formosa"), was one of the major units ordered to move north.

The Inchon landing gave urgency to Chinese preparations to enter the war. Two days after the landing on 15 September 1950, a liaison party was sent to Pyongyang. Meanwhile, Kim Il Sung had asked Stalin for help, including putting pressure on China to send troops. Stalin considered the most acceptable assistance by Chinese armed forces would be in the form of "people's volunteers." In a telegram to Mao on 1 October, Stalin advised: "The Chinese soldiers may be considered as volunteers and of course will be commanded by the Chinese." Mao responded the next day that this was his intention.

For many years Western historians supposed that Lin Piao, a legendary Chinese Communist leader, commanded Chinese forces in Korea. They were wrong. At a 4 October conference in Peiping, Lin Piao argued strongly against sending troops into Korea to fight the Americans and refused to lead the intervention, using the subterfuge of poor health. Lin went off to Moscow for medical treatment and Mao named Peng Dehuai, a tough old revolutionary, to take his place. Peng, born in Hunan province of peasant stock, had emerged as a senior commander in Mao Tse-tung's famed Long March in 1934-1935. Peng arrived in Peiping too late for the 4 October meeting but met the next day with Mao who directed him to be ready to enter Korea by 15 October. On 8 October, Mao officially ordered the creation of the *Chinese People's Volunteers*, which would be the expeditionary element of the *Northeast China Border Defense Force*, with Peng as both military commander and political commissar.

That same day, 8 October, Mao sent his adroit vice-chairman and foreign minister, Zhou Enlai, to the Soviet Union to discuss with Stalin the provision of air assistance and military equipment. Zhou met with Stalin at his Black Sea resort. Stalin was noncommittal and said that he was not yet ready to provide air support. In Manchuria, Peng Dehuai was furious when he learned this. He stormed back to Peiping to meet again with Mao.

Meanwhile Kim Il Sung was pressing for immediate Chinese help.

In early October, Zhou Enlai informed the Indian ambassador in Peiping, Kavalam M. Panikkar, that if the United Nations forces crossed the 38th Parallel, China would send troops to defend North Korea. This warning reached Washington through diplomatic channels in New Delhi and London. Substantiating reports came through Moscow and Stockholm. The warnings were forwarded to MacArthur's GHQ in Tokyo.

On 15 October, the famous Wake Island meeting of President Harry S. Truman with General MacArthur took place. Truman's later blunt, but inadequate, explanation for the conference was "I wanted to have a personal talk with the General." A wary MacArthur perceived the meeting as a presidential ambush primarily designed to reinforce the Democratic Party's chances of success in the upcoming congressional elections. According to MacArthur, the possibility of Chinese intervention came up almost casually. He stated in his *Reminiscences* that the general consensus was that China had no intention of intervening. Truman would later say in his *Memoirs* that the threatened intervention in Korea was a prime reason for the meeting. He wanted MacArthur's "firsthand information and judgment."

What Truman took away from Wake Island was that the war in Korea was won and that the Chinese Communists would not attack. Asked about the chances of Chinese intervention, MacArthur, according to Truman, replied that there was very little chance that the Chinese would come in. At the most they might be able to get fifty or sixty thousand men into Korea, but since they had no air force, "if the Chinese tried to get down to Pyongyang, there would be the greatest slaughter."

MacArthur remembered the conversation quite differently. He would later say that it was a "prevarication" that he had predicted, "that under no circumstances would Chinese Communists enter the war." He characterized his Wake Island view on the possibility of Chinese intervention as "speculative." His own local intelligence, filtered through to him by his long-time G-2, Major General Charles A. Willoughby, USA, told him that large numbers of Chinese troops were massed across the Yalu, but his estimate was that America's virtually unopposed air power would make large-scale intervention impossible.

Four nights after the Wake Island meeting, on 19 October, the Chinese in massive numbers began crossing the Yalu.

General Almond's Ambitions

The entry of Chinese troops in force into North Korea was not picked up by United Nations intelligence, neither visually by aerial reconnaissance nor audibly by intercepts of radio signals. In northeast Korea, Almond continued his advance with great confidence. Almond's over-riding ambition was to beat his rival, General Walker, to the Yalu. His X Corps included two strong U.S. divisions—the 1st Marine and the 7th Infantry—and two Republic of Korea or "ROK" divisions—the Capital and 3d—and there were more troops on the way. With the expected arrival of the U.S. 3d Infantry Division the total would come up to 102,000, about two-thirds as many troops as Walker had in his Eighth Army.

The two ROK divisions, organized into the ROK I Corps, could

From left to right: MajGen David G. Barr, Commanding General, 7th Infantry Division; MajGen Edward M. Almond, Commanding General, X Corps; and Col Herbert B. Powell, Commanding Officer, 17th Infantry. After Barr's division made an unopposed landing at Iwon on 29 October, Almond pushed Barr to get to the Yalu. Powell's regiment was the spearhead for the advance.

Photo by Cpl Alex Klein, National Archives Photo (USA) 111-SC351957

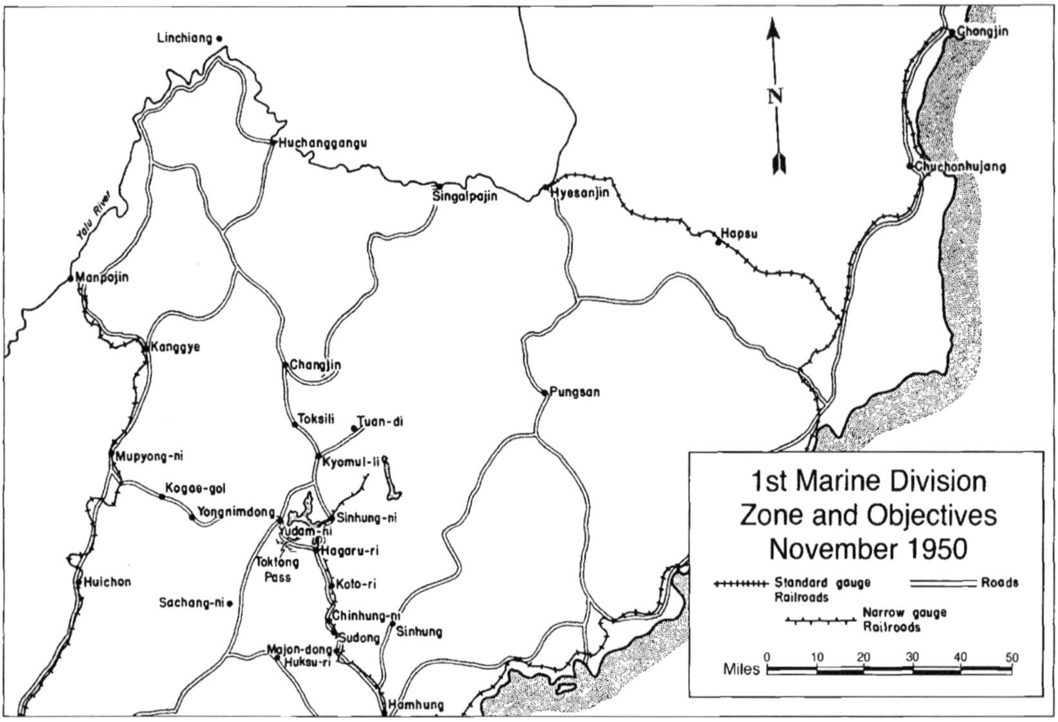

best be described as light infantry. They had no tanks and their only artillery were obsolescent 75mm howitzers. Almond's optimistic assessment of the ROK corps' fighting capabilities was "that they were a good deal better than the people they were chasing, the disorganized, disabled North Korean force."

"I realized," said Almond years later, "that we were scattered all over the landscape, but the general deployment was controlled by the terrain of the area in which the [X] corps was to operate." Almond should also have realized that there were strings tied to his employment of the 1st Marine Division and its companion 1st Marine Aircraft Wing. Shepherd, in Hawaii, was watching the use of the Marines very closely and so was the Commandant of the Marine Corps, General Clifton B. Cates, in Washington.

Army Major General David G. "Dave" Barr's 7th Division had loaded out from Pusan on 19 October as a follow-on to the 1st Marine Division at Wonsan. Barr had been the chief of staff of several commands in Europe during World War II. After the war he had headed the Army Advisory Mission in Nanking. At the war's beginning his 7th Division had been stripped to provide fillers for the 24th and 25th Divisions, the first divisions to be deployed to Korea. The 7th Division, before following the Marines ashore at Inchon, had been hurriedly brought to war strength with untrained South Korean recruits—the so-called KATUSA or "Korean Augmentees to the United States Army."

Now, when the Marine landing at Wonsan was delayed, Barr's destination was changed to Iwon, 75 miles northeast of Hungnam. Iwon was still theoretically in "enemy" territory, but it had good beaches and was known to be free of mines. Barr reloaded the 17th Regimental Combat Team under Colonel Herbert B. Powell into seven LSTs (tank landing ships) to be used in an amphibious assault in the event that the beaches were defended. They were not. As at Wonsan, the South Koreans had already taken the port from the land side. Powell's RCT-17 landed unopposed on 29 October and plunged ahead in a dash for Hyesanjin on the Yalu River. By the end of the month Powell's lead battalion was in a bitter four-day fight with the beaten, but still stubborn, North Koreans at Pungsan. In the days that followed, the remainder of the 7th Division came ashore. The 31st Infantry began landing on 3 November with the mission of moving in on the left flank of the 17th Infantry. The 32d Infantry followed on 4 November and went into bivouac northeast of Hungnam. On 8 November, the 31st Infantry ran into Chinese troops on the slopes of Paek-san, a 7,700-foot peak. In what was the 7th Division's first contact with the Chinese, the regiment reported at least 50 enemy killed.

Almond considered his control over the ROK I Corps to be no different than that he had over the 1st Marine Division and the 7th Infantry Division. Not all would agree, then or now, that his command of the ROK units was that complete. Their more binding orders came from President Syngman Rhee. While the landings at Wonsan and Iwon were still in prospect, the ROK Capital Division had marched steadily up the coast road to Iwon. The ROK 3d Division, meanwhile, had moved northwest from Hamhung toward the Chosin Reservoir.

On the Eighth Army front, the Chinese, whose presence in Korea had been doggedly denied at GHQ Tokyo, had by late October suddenly surfaced in formidable numbers. By the end of the month, the Chinese had defeated the ROK II Corps on the right flank of the Eighth Army to the point of disintegration, exposing the next unit to the left, the U.S. I Corps. General Walker ordered a general withdrawal to the Chungchon River. The Chinese did not pursue but broke off their offensive as suddenly as it began.

Separating the right flank of Eighth Army from X Corps was the Taebaek mountain range, the spine of the Korean peninsula and supposedly impassable to any significant number of troops. East of the Taebaek Mountains things seemed to continue to go well for General Almond and his X Corps. Almond did not appear to be perturbed by General Walker's problems.

Smith's Commanders and Staff

Because of the widely dispersed missions assigned his 1st Marine Division, General Smith had divided his command into regimental combat teams built around his three infantry regiments. RCT-5, under Lieutenant Colonel Raymond L. Murray, was assigned a zone behind Litzenberg's RCT-7. RCT-1, commanded by the legendary "Chesty" Puller, already the holder of four Navy Crosses, would remain for the time being in the vicinity of Wonsan fighting the remnants of one or more broken North Korean divisions struggling to get north.

All three regimental commanders had been successful battalion commanders in World War II. Moreover, Puller had commanded the 1st Marines at Peleliu. Now in this new war, Murray, 37, had brought the 5th Marines to Korea in a pell-mell rush to play a fire-brigade role in the defense of the Pusan Perimeter. Puller, 52, had arrived from Camp Pendleton with the 1st Marines in time for the Inchon landing. Litzenberg had formed the 7th Marines at Camp Pendleton, California, in a matter of days and had gotten to Korea in time to join in the battle for Seoul. Litzenberg was called "Litz the Blitz" by some, but this was more an alliteration—and maybe a little derisive at that—rather than a description of his command style, which tended to be cautious and buttoned-up. Because of his closely cropped prematurely white hair, some of his irreverent young lieutenants, and perhaps a few of his captains and majors often referred him to, as the "Great White Father."

Murray, the junior regimental commander, was simply known as "Ray." Among Marines, who like to argue over such things, Murray's 5th Marines, with the highest percentage of regulars and the longest time in the fight, would probably have rated highest in combat effectiveness. The 1st Marines, with Chesty Puller as its commander, most likely would have rated second. The 7th Marines, last to arrive and with the highest percentage of reserves, still had to prove itself and would have come in third.

Smith had a strong division staff: some members were already serving with the division when he took command and some that he had subsequently asked for. At the outbreak of the war Colonel Alpha L. Bowser, Jr., 40, had just been assigned as Force Inspector, Fleet Marine Force, Pacific. General Smith asked General Shepherd for his services as G-3 of the division. Shepherd assented and Bowser had become Smith's operational right-hand man. Bowser, Naval Academy 1932, had an enviable reputation in the Corps as a Quantico instructor, staff officer, expert in naval gunfire, and artillery battalion commander at Iwo Jima. He said later: "One of the major problems of the entire operation to the north of Wonsan was our open flank to the west. We never established contact with the right flank of the Eighth Army, and we had a great void out there which I at one time estimated to be somewhere in the neighborhood of 85 to 100 miles."

In a 1971 interview by D. Clayton James, noted historian and biographer of MacArthur, Bowser gave his considered opinion of Almond's leadership:

> General Almond was probably one of the most aggressive corps commanders I have ever seen in action. He was aggressive almost to a fault in my estimation. From

National Archives Photo (USMC) 127-N-A5895

For his brilliant staff work as division G-3, Col Alpha L. Bowser would later receive a Legion of Merit from MajGen Smith. Standing to Bowser's right is Col Bankson T. Holcomb, Jr., the division G-2, who received a similar decoration. These awards were made in January 1951.

> the standpoint of his own personal comfort and safety, he never gave it a thought. He was up at the front a great deal. He was what we referred to as a "hard charger."

However, Bowser went on to say:

> I questioned his judgment on many occasions. . . . I think that General Almond pictured this [campaign] in his mind's eye as a sweeping victory that was in his grasp. But he gambled and he lost. . . . A rather vain man in many ways. Ambitious. Could be a very warm personality as a personal friend. If he had one glaring fault, I would say it was inconsistency.

Bowser was not the only star player on Smith's team. Virtually all of the senior members of Smith's general and special staff were combat-tested veterans of considerable reputation.

Brigadier General Edward A. Craig, Smith's assistant division commander, now 54, had been commissioned in 1917, the same year as Shepherd and Smith. During the World War I years, while Smith was in garrison on Guam and Shepherd was winning laurels in France, Craig was fighting a kind of "cowboys and Indians" bush war against bandits in Haiti and Santo Domingo. In World War II he commanded the 9th Marines, first in training on Guadalcanal and then in combat on Bougainville and Guam—for the last he had a Navy Cross. He left the division in the summer of 1950 to command the 1st Provisional Marine Brigade in its adventures in the Pusan Perimeter, but rejoined the parent division in time for the Inchon landing. Like Smith, Craig was tall, slim, and prematurely white-haired. Smith, perhaps recalling his own troubled period as the division's assistant division commander at Peleliu and being virtually ignored by the division commander, Major General William H. Rupertus, used Craig's services wisely and well, particularly as a roaming extension of his own eyes and ears. But Craig would be at home on emergency leave at a critical time in the campaign.

Smith's chief of staff, Colonel Gregon A. Williams, 54, had joined the division at Camp Pendleton in July. Before that he had been chief of staff of Fleet Marine Force, Pacific. A short, erect man, Williams had the reputation of being a "mean SOB." He had had a remarkable—but not unique for an officer of his vintage—career. He had enlisted in San Diego at the outbreak of World War I, but did not get to France, serving, in due time, in Santo Domingo, China,

Col Gregon A. Williams, shown here as a brigadier general, was Smith's chief of staff. Seldom seen in the field by the troops, Williams ran the division staff and headquarters with an iron hand. Like many senior Marine officers, he had had considerable service in China, both before and during World War II.

Department of Defense Photo (USMC) A42833

Haiti, and Nicaragua. As a sergeant he had been in the Dominican *Guardia Nacional* as a local lieutenant. A young Dominican lieutenant, Rafael Leonidas Trujillo, who would be the country's long-time dictator, became Williams' life-long friend. (The writer recalls that at the 1959 New Years reception at the presidential palace, in what was then Ciudad Trujillo, the single ornament on the grand piano in the ballroom was a silver-framed photograph of Colonel Williams.) Service in Nicaragua brought him a Navy Cross. At the beginning of the Pacific War, as an assistant naval attaché in Shanghai, he was taken into custody by the Japanese and held as a prisoner until August 1942 when he was repatriated because of his diplomatic status. Undaunted, he returned to China to the new Nationalist capital in Chungking where he became involved in the support of guerrilla operations. In the summer of 1944 he was sent to the Pacific to take command of the 6th Marines, then involved in mopping-up on Saipan. He was a consummate chief of staff although not much loved by those who had to work for him. He got along famously with contemporaries such as Craig, but he terrified junior officers. Bowser, after some rough spells, said that, "He and I came to a perfect relationship." According to Bowser, Williams "took no guff" from Almond or Almond's chief of staff. Williams stayed close to the command post. Murray recalled seeing him only five or six times during their respective tours in Korea and then only at the division headquarters.

Much more visible to the command than Williams—and much better liked—was the deputy chief of staff, Colonel Edward W. Snedeker, 47. By training a communications officer, Snedeker had been Craig's chief of staff during the fighting by the 1st Provisional Marine Brigade in the Pusan Perimeter. Snedeker, Naval Academy 1926, had distinguished himself in World War II with a Silver Star from Guadalcanal and a Navy Cross from Okinawa where he commanded the 7th Marines. Smith's personal relationship with Snedeker was much closer than it was with Williams.

The G-1 (Personnel), Lieutenant Colonel Harvey S. Walseth, 39, Naval Academy 1935, had served in China before World War II and was a tank officer at Guadalcanal and Iwo Jima.

The G-2 (Intelligence), Colonel Bankson T. Holcomb, Jr., 42, movie-star handsome with his thin, clipped mustache and something of a bon vivant, was an old China hand. The cousin of Thomas Holcomb, who was the Commandant of the Marine Corps during World War II and the first Marine to reach the grade of four-star general, Bankson T. Holcomb, Jr., had graduated from high school in Peiping in 1925 and served two years as an enlisted man before going to the Naval Academy, Class of 1931. He returned to China in 1934 as an assistant to the naval attaché and Chinese language student, followed by two years as a Japanese language student in Tokyo. His speaking and reading ability in both Chinese and Japanese was rated as "excellent." He was at Pearl Harbor as an intelligence officer in December 1941 when the Japanese struck. In 1943 he returned to China once again, this time to operate out of the Nationalist capital of Chungking with Chinese guerrillas.

The G-4 (Logistics) Colonel

In January 1951, Col Edward W. Snedeker would receive a second Legion of Merit from MajGen Smith for his outstanding performance of duty as the division's deputy chief of staff. Earlier, in the Pusan Perimeter, he had been chief of staff of the 1st Provisional Marine Brigade.

National Archives Photo (USMC) 127-N-A5897

Francis M. McAlister, 45 and Naval Academy 1927, had fought as an engineer at Bougainville, Guam, and Okinawa. Like most Marine officers of his generation, he had served in Nicaragua and China.

Walseth, Holcomb, Bowser, and McAlister were the four pillars of the division's general staff. Bowser, the operations officer, and McAlister, the logistics chief, had a particularly close working partnership.

The much larger special staff ranged in grade from second lieutenant to colonel, from Second Lieutenant John M. Patrick, the historical officer, to Colonel James H. Brower, the division artillery officer. Most of the special staff were double-hatted; they also commanded the unit composed of their specialty. Brower commanded the division's artillery regiment, the 11th Marines, with three organic battalions of 105mm howitzers, a battalion of 155mm howitzers, and a battery of 4.5-inch

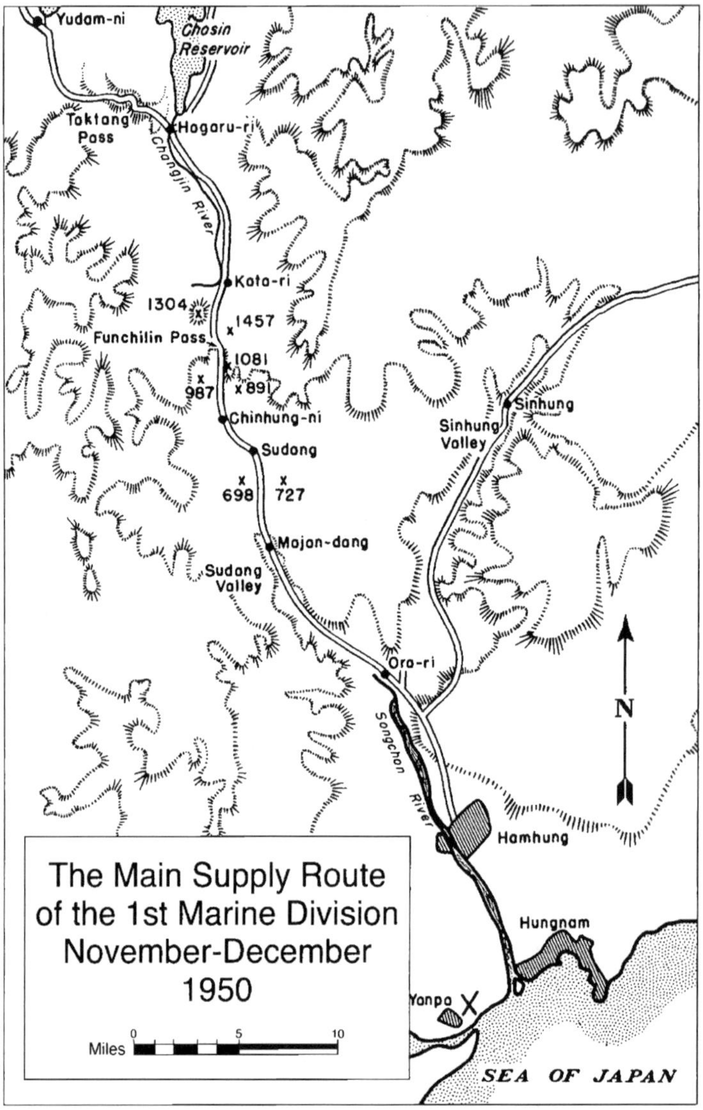

Col James H. Brower was the division artillery officer and commanding officer of the 11th Marines. He fell ill at Hagaru-ri and was replaced in command by his executive officer, LtCol Carl A. Youngdale.
National Archives Photo (USMC) 127-N-A4696

multiple rockets. Brower, 42, was a Virginia Military Institute graduate, Class of 1931, who trained as an artilleryman at Fort Sill. During World War II he served as a staff officer with amphibious forces in the invasions of Sicily and Italy, but arrived in the Pacific in time for Okinawa.

Forty-four-year old Lieutenant Colonel John H. Partridge, as the commander of the 1st Engineer Battalion, was the division engineer. His engineers would work prodigies during the campaign. Partridge, Naval Academy 1936, had been with the 4th Marine Division at Roi-Namur, Saipan, Tinian, and Iwo Jima.

Smith and his staff, and his subordinate commanders and their respective staffs, worked their way through new sets of maps, analyzing the terrain and divining lines of

Distances in Road Miles

Hungnam to Hamhung	8
Hamhung to Oro-ri	8
Oro-ri to Majon-dong	14
Majon-dong to Sudong	7
Sudong to Chinhung-ni	6
Chinhung-ni to Koto-ri	10
Koto-ri to Hagaru-ri	11
Hagaru-ri to Yudam-ni	14
TOTAL	78

communication. Hamhung and Hungnam, then and now, are often confused. Hungnam is the port; it lies on the north side of the Songchon River where it empties into the Japanese Sea. Yonpo, with its airfield, which would prove critical, is on the south side of the estuary. Hamhung is the inland rail and highway nexus, straddling the main rail line from Wonsan north. A narrow-gauge (2' 6") line also started at Hamhung to Chinhung-ni. From there it climbed by cable car, now inoperative, to a plateau in the shadow of the Taebek Mountains. A road paralleled the narrow-gauge line. This would be the main supply route or "MSR" for the Marines' advance. The dirt-and-gravel road stretched 78 miles from Hamhung to Yudam-ni, which, as yet, was just a name on a map. "In only a few weeks," says the Corps' official history, "it would be known to thousands of Marines as the MSR, as if there never had been another."

RCT-7—with the 1st Motor Transport Battalion and Division Reconnaissance Company attached—received a partial issue of cold weather clothing before making the move north by truck and rail during the last three days of October.

Major Henry J. Woessner, 30, Naval Academy 1941, and operations officer of the 7th Marines, was at X Corps command post in Wonsan on 30 October when General Almond, standing before the Corps situation map, briefed General Barr on the upcoming operation. Barr's division was to push north to Hyesanjin on the Yalu. The 1st Marine Division was to reach the border by way of Chinhung-ni, Koto-ri, and Hagaru-ri. After describing this twin-pronged thrust to the Yalu, Almond again turned to the situation map. "When we have cleared all this out," he said with a broad sweep of his hand, "the ROKs will take over, and we will pull our divisions out of Korea."

Before leaving the command post, Woessner talked to an Army liaison officer who had just returned from the ROK 26th Regiment up near Sudong. The Army officer told him that the ROKs had collided with a Chinese force and had been driven back. Colonel Edward H. Forney, Almond's Marine Corps deputy chief of staff, arranged for Woessner to fly over the objective area in an Air Force North American T-6 Texan. Woessner saw no enemy on the flight to and over Hagaru-ri, but the rugged nature of the terrain impressed him.

Woessner, on his return that afternoon to the 7th Marines command post, made his report to Colonel Litzenberg who in turn called in his officers and noncommissioned officers and told them that they might soon be fighting the first battle of World War III. "We can expect to meet Chinese Communist troops," he said, "and it is important that we win the first battle."

Tuesday, 31 October

RCT-7 was scheduled to relieve the ROK 26th Regiment, 3d Division, in the vicinity of Sudong on 2 November. Litzenberg on 31 October cautiously sent out reconnaissance patrols from Hamhung to explore the route northward. One of the patrols from the 1st Battalion, 7th Marines—Captain Myron E. Wilcox, Jr., two lieutenants, three jeeps, and a fire team—reached the command post of the 26th Regiment near Sudong. Wilcox reported to Litzenberg that while there they had seen a Chinese prisoner. The South Koreans told Wilcox and his patrol that they had taken 16 Chinese prisoners and had identified them as belonging to the *124th Chinese Communist Force (CCF) Division*. The prisoners said they had crossed the Yalu in mid-October.

Further interrogation of the 16 prisoners had yielded that they were members of the *370th Regiment* of the *124th CCF Division*, which along with the *125th* and *126th Divisions*, made up the *42d CCF Army*. Roughly speaking, a Chinese army was the equivalent of a U.S. corps. On arriving from the Yalu, the *124th* had deployed in the center to defend the Chosin Reservoir, the *126th* had moved east to the vicinity of the Fusen Reservoir, and the *125th* to the western flank on the

Photo by Cpl Alex Klein, National Archives Photo (USA) 111-SC351718

Marines on 31 October saw for themselves the first Chinese prisoners taken by ROK I Corps. These prisoners, identified as belonging to the 124th CCF Division, *are wearing padded winter uniforms that offered little or no protection for the feet and hands, and the weather was about to turn cold.*

right of the *124th Division*.

Continuing the northward movement of his division from Wonsan, Smith ordered Murray to advance a battalion of the 5th Marines to Chigyong, eight miles southwest of Hamhung. Murray sent his 1st Battalion under Lieutenant Colonel George R. Newton. Newton, 35, Naval Academy 1938, had been a company commander in the Embassy Guard at Peiping in 1941 and spent World War II as a prisoner of war. Now, as a battalion commander, he had done well at Pusan and Inchon. One of Newton's companies was detached to relieve a company of the 7th Marines that was guarding Yonpo Airfield, five miles southwest of Hungnam.

Wednesday, 1 November

As yet the Marines had encountered no enemy anywhere along the MSR from Wonsan north to Hamhung; but Litzenberg was certain that he soon would be facing Chinese adversaries. On the following day, 1 November, he sent a stronger patrol from the attached division Reconnaissance Company to reconnoiter the Huksu-ri area about 45 miles northwest of Hungnam. This patrol, mounted in 21 jeeps and under First Lieutenant Ralph B. Crossman, after running into a small North Korean guerrilla force about three miles short of its objective, dug in for the night.

Meanwhile, the Marines began hearing rumors that the Eighth Army's 1st Cavalry Division—which they had last seen when the 1st Cavalry passed through the Marine lines north of Seoul headed for the successful capture of Pyongyang, the North Korean capital—was in serious trouble. If division headquarters had more

A 3.5-inch rocket section with Company C, 1st Battalion, 7th Marines, holds a position outside Hambung on 31 October. The next day the 7th Marines would begin its march northward to relieve the 26th ROK Regiment near Sudong-ni.

Photo by Cpl Peter W. McDonald, National Archives Photo (USMC) 127-N-A5129

Photo by Cpl Peter W. McDonald, Department of Defense Photo (USMC) A4525

The 7th Marines began its motor march north from Hungnam on 1 November. A cautious Col Homer Litzenberg ordered LtCol Raymond G. Davis, commanding officer of the 1st Battalion, to reconnoiter to his front. Davis' Marines moved forward on foot. A crackle of small arms fire caused the column to halt momentarily. A full issue of cold-weather clothing has not yet been received. Only one Marine is wearing a parka.

definitive information, it did not filter down to the troops. Some 60 miles to the Marines' west in the Eighth Army zone of action, the Chinese had roughly handled the 8th U.S. Cavalry Regiment and the ROK 6th Division during the last days of October, but this had little or no effect on Almond's plans. Litzenberg's orders to advance remained unchanged. His first objective was to be Koto-ri.

During the day on 1 November, the 7th Marines made a motor march from Hungnam to an assembly area behind the ROK 26th Regiment, midway between Oro-ri and Majon-dong, without incident. Nevertheless, a cautious Litzenberg ordered Lieutenant Colonel Raymond G. Davis, 35, Georgia Tech 1938, to make a reconnaissance-in-force to South Korean positions north of Majon-dong with his 1st Battalion, 7th Marines. In World War II, Davis had commanded a heavy weapons company at Guadalcanal and an infantry battalion at Peleliu. The latter battle brought him a Navy Cross.

Late in the afternoon the regimental combat team curled up into a tight perimeter for the night. As part of RCT-7, Litzenberg had Major Francis F. "Fox" Parry's 3d Battalion, 11th Marines; the division Reconnaissance Company under Lieutenant Crossman; Company D, 1st Engineer Battalion, Captain Byron C, Turner; Company E, 1st Medical Battalion, under Lieutenant Commander Charles K. Holloway; detachments from the division's Signal Battalion, Service Battalion, and Military Police Company; and most of the 1st Motor Transport Battalion, Lieutenant Colonel Olin L. Beall.

Thursday, 2 November

The ROK 26th Regiment, awaiting relief, had withdrawn to a position about four miles south of Sudong. Early on the morning of 2 November, the South Koreans were probed by a CCF combat patrol estimated to be about two platoons in strength. Later that morning, Davis' 1st Battalion led the way out of the 7th Marines' perimeter toward the ROK lines at Majon-dong. Major Webb D. "Buzz" Sawyer, 32, followed with the 2d Battalion. A graduate of the University of Toledo, Sawyer had been commissioned in 1941. During World War II, as a captain and major he served with the 4th Marine Division at Roi-Namur, Saipan, Tinian, and Iwo Jima. Afterward, as an instructor at Quantico, he was known as an expert in the reduction of fortified positions.

Corsairs from Marine fighter squadron VMF-312 flew cover for what was essentially a parade northward. The passage of lines with the ROKs was over by 1030. The point, Company A, under Captain David W. Banks, took some scattered long-range fire and suffered a few casualties. Resistance thickened.

Major "Fox" Parry, 32, commanding the 3d Battalion, 11th Marines, was a Naval Academy graduate, Class of 1941. He had been the executive officer of an artillery battalion during Okinawa and immediately before Korea he had taken the yearlong Advanced Artillery Course at Fort Sill. At noon Battery I of Parry's artillery battalion fired the first of 26 fire missions covering the advance that would be shot during the day.

VMF-312, commanded by Lieutenant Colonel J. Frank Cole, flew 12 close air support missions and, as the light failed, night-fighter squadron VMF(N)-513, under Major J. Hunter Reinburg, delivered a few more.

Both Cole and Reinburg were experienced squadron commanders. Cole, 35, had entered the

Photo by Cpl Peter W. McDonald, Department of Defense Photo (USMC) A4549

Guns of Maj Francis F. "Fox" Parry's 2d Battalion, 11th Marines, covered the advance of the 7th Marines to Sudong. Here a 105mm howitzer from Capt Samuel A. Hannah's Battery G, laid for high-angle fire, awaits a fire mission. First artillery missions, shot on 2 November, alternated with close air support strikes by Marine Corsairs.

Marine Corps from the University of Nebraska in 1939 and had commanded fighter squadron VMF-111 in the Central Pacific and VMF-312 at Okinawa. Reinburg, 32, was the stepson of Marine Corps aviation great Lieutenant General Clayton C. Jerome. Reinburg had enlisted in the Naval Reserve in 1936 and transferred to the Marine Corps as an aviation cadet in 1940. Flying as a captain in the Solomons he became an ace, shooting down seven Japanese planes and destroying seven more on the ground. Before taking command of VMF(N)-513 he had spent a year as an exchange pilot flying night fighters with the British Royal Air Force. Technically, Marine fighter-bomber squadrons designated as "VMF(N)" were all-weather squadrons, but the "N" universally caused them to be called "night fighters." Reinburg's squadron flew twin-engine Grumman F7F-3Ns Tigercats

The main body of RCT-7, moving along the road in what Litzenberg called a "walking perimeter," had advanced just short of a mile by nightfall. Davis' 1st Battalion's nighttime positions were less than a mile south of Sudong, stretching across the valley from high ground to high ground. Behind him was Sawyer 's battalion similarly disposed. Sawyer was responsible for the high ground on both sides of the line of march. Captain Milton A. Hull, 30, commanding Company D, had some problems going up Hill 698 on the left hand side of the road. (Hills and mountains—both always called "hills"—were designated by their height in meters above sea level. Thus Hill 698 would be 698 meters or 2,290 feet in height.) A ROK company had precipitously given up its hillside position. The South Koreans, as they passed hurriedly southward, pointed back over their shoulders exclaiming "Chinese!"

Hull, a University of Florida graduate, had been commissioned in 1942 and had spent a good part of the war in China with the guerrillas, possibly with some of the same Chinese soldiers he was now fighting. Easy Company, under Captain Walter D. Phillips, Jr., passed through Hull's Dog Company to complete the fight, getting almost to the crest just

Col Homer Litzenberg called his road march a "walking perimeter." Here a part of the column pauses off the road, while Marine artillery and air pound the hills ahead. At nightfall on 2 November, LtCol Ramond Davis' 1st Battalion, lead element of the main body, halted one mile short of Sudong.

Department of Defense Photo (USMC) A4498

Photo by Sgt Frank C. Kerr, Department of Defense Photo (USMC) A4438

These Marines are members of Company D, 2d Battalion, 5th Marines. By 2 November, LtCol Raymond L. Murray's regiment had moved up by train to Hamhung with orders to patrol between Hamhung and Chigyong. Two days later the 3d Battalion was positioned near Oro-ri and the 2d Battalion was sent into Sinhung Valley. The 1st Battalion remained at Chigyong.

before midnight. (Rifle companies were almost invariably called by their name in the phonetic alphabet of the time: "Dog," "Easy," "Fox," and so on.) Farther to the rear was Major Maurice E. Roach's 3d Battalion, in a perimeter of its own, protecting the regimental train.

That morning, 2 November, Smith had met again with Almond. The 2d and 3d Battalions of the 5th Marines were moving by train to Hamhung. The 1st Battalion had already gone northward. Smith pointed out that the main supply route from Wonsan would be left exposed to guerrilla attack. Almond was not disturbed. He said that patrols could handle the guerrilla situation. Puller's 1st Marines, supported by elements of the 1st Tank Battalion, was given the responsibility from Wonsan northward to as far as Munchon. Murray's 5th Marines would patrol south from Hungnam to Chigyong. This left 54 miles from Chigyong south to Munchon uncovered except for light patrolling by Almond's Special Operations Company and a handful of South Korean counterintelligence agents.

Puller returned to Wonsan at midday on 2 November with his 1st Battalion. The 2d Battalion came back from Kojo-ri the following day. The 3d Battalion, 1st Marines, was still heavily engaged at Majon-ni, 26 miles west of Wonsan.

Friday, 3 November

Litzenberg did not know it, but he was two-thirds surrounded by the *124th CCF Division*. The *371st Regiment* was in the hills to his north and west. The *370th Regiment* was to his east. Somewhere behind these assault regiments, the *372d Regiment* stood ready in reserve.

By midnight on 2 November the 1st and 2d Battalions of the 7th Marines were being probed. An hour later both battalions were bending back from the weight of assaults on their flanks and Marines became acquainted with the Chinese habit of using flares

By 3 November, the 7th Marines was surrounded on three sides by the Chinese 124th Division. *Fighting grew fierce and casualties mounted. A sturdy masonry building in the shadow of high-tension lines coming down from the hydroelectric plant on the Changjin plateau became a battalion aid station.*

Photo by Cpl Peter W. McDonald, National Archives Photo (USMC) 127-N-A4541

National Archives Photo (USMC) 127-N-A4489
MajGen Smith used his assistant division commander, BGen Edward A. Craig, as his roving eyes and ears. Here, on 3 November, Craig, left, checks with Col Litzenberg whose 7th Marines had just made its first solid contact with the Chinese. Litzenberg is digging away at a half-size can of C-ration, probably canned fruit, a favorite.

Department of Defense Photo (USMC) A4501
Chinese prisoners, taken by Col Litzenberg's 7th Marines in the 3-4 November fighting, were identified as being members of the 370th *and* 371st Regiments *of the* 124th Division. *Chinese Communist "volunteers" fought well, but were surprisingly docile and uncomplaining once captured.*

and bugle calls to signal their attacks.

On the MSR, the roadblock in front of Able Company let a T-34 tank go by, thinking it was a friendly bulldozer. The single tank pushed through the company headquarters area and on through the battalion's 81mm mortar position, reaching Davis' command post. The startled Marines engaged the tank with rocket launchers and recoilless rifles; the tank took one or two hits and then turned around and headed north.

All three of Davis' rifle companies suffered heavy casualties as the night went on. The Chinese attackers got down to the road and wedged their way between the 2d and 3d Battalions. The regiment's 4.2-inch Mortar Company was overrun and lost one of its tubes. When morning came a confused situation faced the Marines. The Chinese were still in the valley. Getting rid of them would be an all-day effort. At first light Cole's VMF-312 came overhead with its Corsairs and was joined in mid-morning by Reinburg's Tigercats, pounding away with rockets, fragmentation bombs, and cannon fire. Parry's howitzers rendered yeoman service; before the end of the day his 18 guns had fired 49 missions delivering 1,431 105mm rounds. At closer range, Marine riflemen flushed out the Chinese enemy, fragmented now into individuals and small groups.

This would be Reinburg's last show. On 4 November he relinquished command of VMF(N)-513 to Ohio-born Lieutenant Colonel David C. Wolfe, 33, Naval Academy, Class of 1940. A big, athletic man, Wolfe had taken flight training as a captain and had commanded scout-bomber squadron VMB-433 in the Southwest Pacific during World War II.

The 1st Battalion counted 662 enemy dead in its zone of action. The 2d Battalion did not make a precise count but could not have been far behind. When Marine trucks came up with resupply, they carried back to Hungnam about 100 wounded Marines. Total Marine casualties for the two days—2 and 3 November—were 44 killed, 5 died of wounds, 1 missing, and 162 wounded, most of them in the 7th Marines.

As recorded in the official history by Lynn Montross and Nicholas A. Canzona, a tactical principle was emerging: "To nullify Chinese night attacks, regardless of large-scale penetrations and infiltration, defending units had only to maintain position until daybreak. With observation restored, Marine firepower invariably would melt down the Chinese mass to impotency." It was a principle that would serve the Marines well, time after time, in the coming several weeks.

Photo by Cpl L. B. Snyder, National Archives Photo (USMC) 127-N-A4463

BGen Edward Craig, on the left and wearing a Marine "fore-and-aft" cap, and MajGen Edward Almond, on the right with map case and wearing an Army cap, went forward to the 7th Marines command post to see for themselves the Chinese prisoners that had been taken. Col Homer Litzenberg, the regimental commander, is talking with BGen Craig.

Saturday, 4 November

The 7th Marines' positions remained essentially the same during the night of 3-4 November. The perimeters were peppered lightly, but there were no further Chinese assaults. Later it was learned that the *370th* and *371st CCF Regiments* were withdrawing to a defensive line, established by the *372d Regiment* about two miles north of Chinhung-ni, stretching from Hill 987 to Hill 891.

Wounded Marines arrive at the battalion aid station of Maj Webb D. "Buzz" Sawyer's 2d Battalion, 7th Marines. Helicopter evacuation was still the exception rather than the rule. Most wounded Marines were hand-carried down to the closest road and then moved by jeep to the nearest aid station. Here they would be sorted out ("triage") and sent to the rear for more definitive treatment.

Photo by Cpl Peter W. McDonald, National Archives Photo (USMC) 127-N-A4546

Litzenberg ordered increased patrolling to the north to begin at dawn on 4 November. Marines from Davis' 1st Battalion patrolled to the edge of Sudong, met no resistance, and returned to their perimeter. Crossman's Reconnaissance Company moved out in its jeeps at 0800. First Lieutenant Ernest C. Hargett took the point into Sudong and met a party of Chinese in the middle of the town. Hargett's men killed three and took 20 more as willing prisoners. Crossman now put Second Lieutenant Donald W. Sharon's 2d platoon into the point with the 1st Battalion coming behind them into Sudong.

The North Korean People's Army (NKPA) skeleton *344th Tank Regiment*, down to five Soviet-built T-34 tanks and apparently unable to negotiate Funchilin Pass, had been left on the low ground to fend for itself. One T-34 was abandoned after being damaged in its wild one-tank attack against the 7th Marines command post. The remaining four tanks took covered positions off the road. Sharon passed by the first hidden T-34 but bumped into the second. He and two of his Marines damaged the tank with hand grenades. Charlie Company, 7th Marines, with its own 3.5-inch rockets and reinforced with a section of 75mm recoilless rifles, came on the scene and finished off the second tank. A third tank emerged from a thatched hut. Engaged by both rocket launchers and recoilless rifles, the tank continued to move until stopped by the 5-inch rockets of a flight of Corsairs. The Marines now found the bypassed first tank.

Photo by Cpl Peter W. McDonald National Archives Photo (USMC) 127-N-A4496

Bodies of dead Marines, covered with ponchos and shelter halves, await further disposition. The 7th Marines fight at Sudong lasted three days, 2-4 November. Division casualties for the period, nearly all of them in the 7th Marines, totaled 61 killed in action, 9 died of wounds, 162 wounded, and 1 missing in action.

After receiving fire, the crew surrendered their tank and themselves. The fourth tank, now alone, surrendered without a fight. The *344th NKPA Tank Regiment* was no more. Litzenberg, having advanced almost four miles by mid-afternoon, ordered his regiment to halt for the night in a tight perimeter at Chinhung-ni.

For the first 43 miles north from Hungnam the 1st Marine Division's MSR was a two-lane highway passing through relatively flat terrain. At Chinhung-ni the road narrowed to one lane as it went up Funchilin Pass, climbing 2,500 feet in eight miles of zigzagging single-lane road clinging to the sides of the mountains; "a cliff on one side and a chasm on the other" as the official history described it. The narrow gauge railroad was operable as far as Chinhung-ni and it was decided to establish a railhead there.

The division Reconnaissance Company was ordered to move forward another mile, on up into Funchilin Pass, and outpost the southern tip of Hill 891. With Second Lieutenant Charles R. Puckett's 3d Platoon out in front, the reconnaissance Marines moved almost into the saddle separating Hills 987 and 891, already inconveniently occupied by the Chinese. A firefight developed. The company held its ground but lost two Marines killed, five wounded, and two jeeps destroyed.

On 4 November, Smith shifted his command post from Wonsan to Hungnam, occupying an abandoned engineering college on the outskirts of the city. In reconnoitering for the site, Smith's assistant division commander, Brigadier General Craig had been treated to the sight of 200 dead Koreans laid out in a row, executed by the Communists for no apparent reason. Smith flew to Hungnam by helicopter and occupied the new command post at about 1100. Most of his headquarters arrived by rail that evening, an uneventful trip except for a few scattered rifle

The 7th Marines entered Sudong on 4 November. Beyond Sudong the main supply route began its climb into Funchilin Pass. Here, a Marine patrol, troubled by a sniper, searches out a hamlet of thatched-roofed, mud-wattle huts.

Department of Defense Photo (USMC) A4537

Sketch by Sgt Ralph Schofield, USMCR

A Marine sniper draws a bead on a distant Chinese enemy. This sketch, and numerous others that follow, are by Cpl (later Sgt) Ralph H. Schofield, a talented Marine Corps reservist from Salt Lake City, who served as a Leatherneck *magazine combat artist. A seasoned veteran of World War II, Schofield had fought as an infantryman in the South Pacific.*

shots. The larger part of his headquarters would remain in place in Hungnam for the duration of the operation.

To the south of the 7th Marines, the battalions of Murray's RCT-5 were having their own adventures. "Our first assignment was to go to the east side of the reservoir," remembered Murray. "I wondered, why are they splitting us up like this?" By 4 November, the 1st Battalion (Lieutenant Colonel George R. Newton) had been left behind at Chigyong and detached to division control. The 3d Battalion (Lieutenant Colonel Robert D. "Tap" Taplett) was positioned near Oro-ri. The 2d Battalion (Lieutenant Colonel Harold S. Roise) had been sent into Sinhung Valley, five miles north and 15 miles east of the 7th Marines, to relieve the ROK 18th Regiment. The relief was accomplished without incident. Roise, 34, from Idaho, had spent World War II in the battleships *Maryland* (BB 46) and *Alabama* (BB 60). The 2d Battalion, 5th Marines, was his first infantry command and he had done well with it at Pusan, Inchon, and Seoul. His mission now was to block the Sinhung corridor and to find a northerly route to either the Chosin Reservoir or to the reservoir known to the Marines by its Japanese name "Fusen." The Korean name was "Pujon." Roise's mission carried him away from 1st Marine Division's axis of advance and into the zone of the 7th Infantry Division.

With 7th Marines, 5-6 November

Early on Sunday morning, 5 November, the Major Roach's 3d Battalion, 7th Marines, passed through the 1st Battalion to con-

SSgt Meyer Rossum triumphantly displays a poster of Soviet dictator Joseph Stalin found in a by-passed Chinese bunker in the vicinity of Funchilin Pass. Marines would learn that the Chinese, despite problems of weather and terrain, were avid diggers and experts at field fortification.

National Archives Photo (USMC) 127-N-A4807

tinue the advance up Funchilin Pass. Hargett's reconnaissance platoon led the way. Rounding a hairpin turn, Hargett ran into Chinese fire and had four more Marines wounded. The 3d Battalion moved into the attack. Item Company was given Hill 987 and George Company Hill 891 as their objectives. Both were stopped by mid-morning by heavy small arms and machine gun fire. For the rest of the day the battle continued as a duel between Parry's 105mm howitzers and Chinese 120mm heavy mortars. From overhead, the Corsairs of VMF-312 delivered 37 close air support sorties.

At the top of the pass the road flattened onto a plateau and ran for two miles until it reached the village of Koto-ri where it rejoined the now-abandoned narrow gauge railroad. During the day General Smith gave Litzenberg the objective of reaching Koto-ri.

Roach's 3d Battalion continued the attack the next morning. How Company, under First Lieutenant Howard H. Harris, was to pass through George Company and move up the southern tip of Hill 891. Item Company, under First Lieutenant William E. Johnson, was to continue its attack against Hill 987. Both attacks went slowly, with the assaults not getting underway until mid-afternoon.

Second Lieutenant Robert D. Reem, leading one of How Company's platoons in the final assault, threw himself on a Chinese grenade and was killed. Harris radioed Roach that his company was exhausted. Roach relayed the report to Litzenberg who ordered the company to disengage and withdraw. Next morning, 7 November, Roach's battalion again moved up the slopes of both Hills 891 and 987, and this time found them empty of enemy. The Chinese had disappeared during the night. For most of the next three weeks traffic northward on the MSR would be unimpeded.

Operations North of Wonsan, 4-9 November

Meanwhile, in accordance with Almond's decision on 3 November, X Corps troops and the 1st Marine Division continued to share the responsibility for the Wonsan-Hungnam MSR. Operation of the Wonsan-Hamhung rail line came under X Corps Railway Transportation Section. The division began sending supply trains north daily from Wonsan. For two days they got through unmolested, but on the third day, 6 November, the train was halted at Kowan by torn-up rails. North Korean guerrillas then attacked the train, which was guarded by 39 Marines from Charlie Company of the 1st Amphibian Tractor Battalion. Taken by surprise, eight Marines were killed, two wounded, at the outset. Six more Marines were wounded in the ensuing firefight. The guard then broke off action and found protection within the perimeter of an Army artillery battalion.

Smith was promised the use of the Army's newly arrived 65th Regimental Combat Team to guard bridges and other key points along the route. Rail service from Wonsan to Hamhung was resumed on 9 November with the caution that passengers were to ride only in open gondola cars. Their steel sides promised some order of protection from small arms fire and mortar fragments. While Marines rattled northward in gondola cars, MacArthur on 9 November informed the Joint Chiefs of Staff that complete victory was still possible and reiterated his belief that U.S. air power would prevent the Chinese from crossing the Yalu in decisive numbers.

Arrival of 3d Infantry Division

The Army's 3d Infantry Division, its ranks hastily filled out with South Koreans, began arriving at Wonsan in early November. Major General Robert H. "Shorty" Soule, the division commander, was a paratrooper who had fought with the 11th Airborne Division under MacArthur in the Southwest Pacific. The first regiment of Soule's division to land was the 65th Infantry, made up largely of Puerto Ricans, on 5 November. Almond came, looked, and said he "didn't have much confidence in these colored troops."

During World War II, Almond had commanded the U.S. 92d Infantry Division which had almost all white officers but black rank-and-file. The division had turned in a mixed performance in Italy. Almond's prejudices were typical of his generation and Southern background. The regimental commander, Colonel William W. Harris, West Point 1930, protested that most of his men were not "colored," but "white." Almond, unconvinced of the 65th RCT's reliability, told Harris that he was going to send the regiment north to Yonghung and then west across the mountains to make contact with the Eighth Army's right flank. Harris was appalled by these orders.

The 1st Shore Party Battalion, under command of legendary Lieutenant Colonel Henry P. "Jim" Crowe, 51, stayed behind at Wonsan to help the 3d Infantry Division land and unload. Crowe had enlisted during World War I and had a fabled career as football player, team shot, and bandit fighter, reaching the highly prized warrant grade of Marine gunner in

1934. He had a Silver Star from Guadalcanal and a Navy Cross from Tarawa where he commanded a battalion as a major. He thought Soule "one of the finest men" he ever met, but he found Almond "haughty."

5th Marines Operations, 5-8 November

There was now clear evidence that the Chinese, and some North Koreans, were out in front of Roise's 2d Battalion, 5th Marines, in Sinhung Valley but keeping their distance. Northwest of Sinhung itself and about 10 miles due east of Koto-ri, Dog Company captured a stray Chinese soldier found sleeping in a house. He proved to be a wealth of information. He said that he belonged to the *126th CCF Division*. He asserted that six CCF armies had arrived in North Korea and that a total of 24 divisions had been committed to the intervention. He had learned this in a series of lectures given by political officers to his regiment after it had crossed the border.

Smith conferred with Almond on the afternoon of 7 November. "He apparently has been somewhat sobered by the situation on the 8th Army front, which is not very good," Smith entered into his log. Almond promised Smith that he would let him concentrate the 1st Marine Division.

The 1st Battalion, 5th Marines, remained at Chigyong. On 7 November, Major Merlin R. Olson, 32, the battalion's executive officer, with Companies A and B reconnoitered in force west of Oro-ri to Huksu-ri. On the 8th, still short of his objective, Olson ran into a North Korean force, estimated at 2,000, and was recalled.

Meanwhile, Roise's patrols had found no useable road to either Chosin or Fusen Reservoirs but had learned that the road leading northeast to the Manchurian border, into the zone assigned to the U.S. 7th Infantry Division, could bear military traffic. One of his patrols touched a patrol from the 31st Infantry on 8 November. Smith had an understanding with Almond that if the 5th Marines could not get to the Fusen Reservoir by road, Barr's 7th Infantry Division would attempt to reach it from the east.

Photo by Cpl L. B. Snyder, National Archives Photo (USMC) 127-N-A4620

LtCol Raymond Davis' 1st Battalion, 7th Marines, occupied Koto-ri on the Marine Corps Birthday, 10 November, against no resistance. A day later needed replacements arrived. One or two battalion-sized replacement drafts arrived in Korea each month to keep the division's ranks—particularly the infantry units—at fighting strength.

7th Marines Operations, 8-11 November

On 8 November General Almond visited the 7th Marines. On learning that Captain Thomas E. Cooney, commander of George Company, had been twice slightly wounded on Hill 891, he awarded Cooney an on-the-spot Silver Star. His aide was caught without a supply of medals. Almond scribbled a note on a piece of paper—"Silver Star for Gallantry in Action"—and pinned it to Cooney's jacket.

A patrol of 15 Marines under First Lieutenant William F. Goggin of the 2d Battalion left Chinhung-ni at noon on 8 November, reached Koto-ri, and next evening returned unscathed to the lines of the 3d Battalion. Next day, 10 November and the Marine Corps Birthday, the 1st Battalion passed through the 3d Battalion and an hour-and-a-half later entered Koto-ri.

X Corps issued an order attaching the 65th Infantry and the ROK 26th Regiment to the 1st Marine Division. Two battalions of South Korean Marines were also to be attached. On receiving the order Smith learned that he was responsible for making contact with the Eighth Army. He gave orders to that effect to the 65th Infantry and was annoyed to find that Almond had already given the regiment's commander, Colonel Harris, detailed instructions down to the company level as to what to do. Something of the same happened with regards to the mission of the ROK 26th Regiment. "Such a procedure, of course, only creates confusion," Smith fussed in his log. "It was this type of procedure

Photo by Sgt John Babyak, National Archives Photo (USMC) 127-N-A4562

At division headquarters at Hamhung, MajGen Smith observed the Marine Corps Birthday in traditional fashion. He read the birthday message from the Marine Corps Manual *and then cut the somewhat meager cake with a Korean sword. As tradition prescribes, the first slice went to the oldest Marine present, BGen Craig.*

Koto-ri, a hapless little hamlet. As the official history observed, the cold seemed "to numb the spirit as well as the flesh." On the 11th, Company C, 1st Battalion, 7th Marines, had a fight in which it lost four Marines killed, four wounded, and claimed 40 enemy casualties. Otherwise the enemy seemed to have vanished.

5th Marines Operations, 9-13 November

Murray received orders on 9 November to concentrate his regiment on the MSR leading to Chosin Reservoir. Newton's 1st Battalion, coming out of Chigyong on 10 November, was to move to Majon-dong. A patrol sent forward from Newton's battalion was ambushed and had to be rescued with a battalion-sized attack before the battalion could get to the village. On the 13th, another patrol from the 1st Battalion, 5th

which I protested to General Almond in connection with direct orders given to my regiments." To Smith's further annoyance, he was ordered to provide a rifle company to guard X Corps command post at Hamhung. The order was passed to the 3d Battalion, 5th Marines, for execution. Taplett received it with "surprise and disgust," not understanding why a headquarters with about 2,000 troops needed extra security so far behind the lines of advance. He detailed Item Company under Captain Harold G. Schrier to do the job—the same Schrier who as a lieutenant had taken his platoon up Mount Suribachi on Iwo Jima to raise the first flag.

That evening there was a Marine Corps Birthday party in General Smith's mess attended by his staff. Punch and cake were served. Smith entered in his log: "I read the paragraphs from the Marine Corps Manual and then cut the cake with a Korean sword."

The weather had turned terrifically cold up on the plateau, well below zero at night. Platoon warming tents were set up in

Col "Chesty" Puller cuts the Marine Corps Birthday cake on 10 November at his 1st Marines regimental headquarters outside Wonsan where the weather was still pleasant. Far to the north, on the Chosin plateau, the 7th Marines was already encountering sub-zero temperatures.

Photo by Cpl W. T. Wolfe, National Archives Photo (USMC) 127-N-A4571

Department of Defense Photo (USMC) A4628

On 12 November the villagers at Koto-ri were informed that they have been "liberated" and were now free to elect their own village officials. The large number of Korean civilians, who would later crowd into the Marines' defensive perimeters, would become a huge problem.

Marines, ran into a company-sized group of Chinese that killed seven Marines and wounded three more before withdrawing.

Roise's 2d Battalion came out of Sinhung Valley on 13 November with orders to relieve the 7th Marines of the responsibility of defending Koto-ri. Along the way Roise's Marines picked up one Chinese and 12 North Korean prisoners. An airstrip capable of handling light aircraft was opened at Koto-ri that same day. Taplett's 3d Battalion, 5th Marines, was now at Chinhung-ni. Taplett, 32, from South Dakota, had spent most of World War II on sea duty with the cruiser *Salt Lake City* (CL 25). The 3d Battalion was his first infantry command. He was not always an easy personality, but his performance at Pusan, Inchon (where he had led this battalion ashore in the successful seizure of Wolmi-do in the opening phase of the landing), and Seoul had been outstanding. Still stiffer fights were ahead of him.

Smith, making his own road reconnaissance of Funchilin Pass, took a helicopter as far as Chinhung-ni. Helicopters at that time, because of the cold and altitude, were not going farther north; there being problems with gear boxes freezing up. Smith borrowed a jeep from Taplett and drove on up to Koto-ri.

MacArthur Reassesses Situation

By now MacArthur had to accept that the Chinese were in Korea in strength, perhaps as many as 100,000 of them, but he was still of the opinion that China would not make a full-scale intervention.

Almond had moved his headquarters on 11 November from Wonsan to Hamhung with plans to move his command post farther north to Hagaru-ri. Almond must have reflected that 11 November was Armistice Day from the First World War. Many of the senior leaders in Korea had fought in that war, including MacArthur as a brigadier general and Almond as a major. Almond had served with distinction in the U.S. 4th Infantry Division as commander of the 12th Machine Gun Battalion. Armistice

(Continued on page 34)

The wreath with "Merry Christmas," perhaps some Marine's idea of humor, is misleading. The photo was probably taken at Chinhung-ni in mid-November. At left is LtCol Robert D. Taplett, Commanding Officer, 3d Battalion, 5th Marines, and at right is his executive officer, Maj John J. Canny, who would die at Yudam-ni two weeks later.

Photo by Sgt Frank C. Kerr, National Archives Photo (USMC) 127-N-A7349

Coping with the Cold

How cold was it at the reservoir? Most Marines "knew" that the temperature went down to about 25 degrees below zero at night, but how many Marines had a thermometer in their pack?

The cold was no great surprise, unless, perhaps, you were like one Marine from Samoa who had never seen snow before. The division staff knew by late October or early November that Hagaru-ri had the reputation of being the coldest place in North Korea, with a recorded temperature of 35 degrees below zero. The climate is roughly like that of Minnesota or North Dakota. The winter of 1950 was a cold one, but not unusually so. The powers that be had adequate warning that it was coming and considerable preparations had been made.

Those at the top, and some at other levels in the 1st Marine Division, had had some experience with cold weather operations, if not by participation, at least by observation and a bit of training.

The division's commanding general, Major General Oliver P. Smith, had gone with the 1st Provisional Marine Brigade to Iceland in August 1941 to relieve the British garrison, as a major in command of the 1st Battalion, 6th Marines. He remembered Iceland as a "bleak and rugged island—mountains, cliffs, no trees—not a tree" and most of all the violent, never ceasing wind.

National Archives Photo (USN) 80-G-424584

These Marines, looking very fresh, pause enroute to Yudam-ni on 27 November to heat their C-rations over an open fire. They are wearing the newly issued shoe-pacs with boot socks folded neatly over the tops. At the far left the Marine appears to be wearing old-fashioned galoshes over field shoes or boots, a better combination against the cold than the shoe-pacs.

Cold was all-pervasive, even at Hamhung, which with its near sea-level elevation was much milder than up on the Changjin plateau. Here Marines at Hamhung, probably members of a combat service unit, cook bacon and beans on top of a stove made from a gasoline drum.

Department of Defense Photo (USMC) A4617

There were others, besides Smith, in the division who had also been in the Iceland expedition. One of them was Lieutenant Colonel Raymond L. Murray, commander of the 5th Marines. In Iceland he had been a captain and commander of a machine gun company. He had also served in Peiping in North China before World War II. He did not find Iceland as rugged as Korea: "It was not terribly cold. I don't think it ever got much below 10 above zero."

The Marines suddenly on their way to Iceland did not, at first, have any specialized winter clothing. They wore their wool kersey winter service uniforms including their woolen overcoats, supplemented by some items bought on the open market, notably some short, sheepskin-lined, canvas coats purchased from Sears Roebuck and carried as organizational property. Another much-favored addition were pile-lined hats with ear flaps, such as Marines had worn in North China.

The Marines in Iceland did not live in tents or in

the open. They were billeted in Nissen huts, "an elongated igloo covered with corrugated iron roofing and lined with beaver board," the flimsy British equivalent of the more substantial American Quonset huts. Marines piled sod on the sides of the Nissen huts to improve insulation. Each battalion had a different camp in a different part of the island.

In many ways the deployment, as an opportunity for winter training, was a disappointment. Finnish success with ski troops in the Winter War with the Soviet Union in 1939 had been much publicized (and romanticized). But it did not get as cold in Iceland as it was supposed to get and there was not much snow, seldom as much as a foot.

Marine experiments with skis and more work-a-day snowshoes did not come to much. Nine years later, the Marines at Chosin Reservoir did not have skis or snowshoes and it was just as well. They would not have been useful.

The brigade came back in February and March 1942 wearing the British Polar Bear shoulder patch—and were ordered to take it off. Most of the Marines would soon be on their way to Guadalcanal, and beyond that to Tarawa, and would earn another shoulder patch, either that of the 1st Marine Division or 2d Marine Division.

A larger percentage of Marines in the division than those few who had been in Iceland were those who had served in North China after the end of World War II, a now almost forgotten episode. It began with the 55,000-man deployment of the III Amphibious Corps at the end of September 1945 that included both the 1st and 6th Marine Division and the 1st Marine Aircraft Wing, and ended with the withdrawal of the last battalion in 1949, just a year before the Marines went to Korea. Like Iceland, it was largely a garrison experience, but the China Marines learned about the sub-zero temperatures and the arctic winds that came out of Manchuria and across the Gobi Desert. Marines guarding supply points or critical bridges or riding the coal trains knew how cold it could get.

The clothing they wore, including the Navy parka, was not much different than that which would be worn in Korea. Officially designated as the Marine Corps' 1943 cold weather uniform, it was predicated largely on the Iceland experience and consisted primarily, except for the parka, of U.S. Army components.

Other, older Marines in the division, including the chief of staff, Colonel Gregon A. Williams, the G-2, Colonel Bankson T. Holcomb, Jr., and the commanding officer of the 1st Marines, Colonel Lewis "Chesty" Puller, had had substantial service in pre-World War II China, including a chance to observe operations by Chinese Communist forces in the cold. They knew about the padded Chinese winter uniforms. Some, including Chesty Puller who was much better read and more of a student of military history than his flamboyant reputation would suggest, had studied Japanese winter operations in northern Korea and Manchuria in the Russo-Japanese War of 1904-05.

The Quartermaster General of the Marine Corps, Major General William P. T. Hill, himself an old China hand and an explorer of the Gobi Desert, began shipping out cold-weather clothing, including Navy parkas, to Korea in October 1950. Beginning in November, the battalion-sized replacement drafts being sent to restore combat losses received rudimentary cold weather training, at least in the wearing of cold-weather clothing.

The Marines, and, for that matter, the U.S. Army, used the "layer principle" for winter clothing, which simply meant that the Marine or soldier piled on as many layers of clothing as he could find. From the skin out he might have on cotton underpants and shirt or "skivvies," winter underwear or "long johns," mustard-colored flannel shirt, utility trousers or green kersey service trousers if he had them, sweater, green sateen winter trousers, alpaca vest, utility coat, a woolen muffler, and perhaps an M1943 field jacket, all crammed under a long, hooded, pile-lined Navy parka. The parka was warm but heavy and clumsy. Some Marines managed to find the shorter anorak-type parka worn by the Army and liked it better. Also popular, when they could be found, were the Army's "trooper" style pile-lined winter hats with earflaps. Several styles of gloves were issued. The most common had a leather and fabric outer shell and an inner mitten of knitted wool.

On their feet, Marines, unless they could find a substitute, wore "shoe-pacs"—waterproof rubber bottoms with laced leather uppers. They were issued with two sets each of felt innersoles and heavy woolen boot socks. The Marines were told to keep one set of the socks and innersoles inside their clothing next to their body and to change them frequently. These instructions were good in theory but difficult to follow in practice. Excessive perspiration, generated by marching, soaked the

National Archives Photo (USMC) 127-N-A5351

When possible, as here at Koto-ri, sleeping holes were dug behind fighting holes and frequently covered with ponchos or shelter halves. In this case, the occupants have managed to incorporate a stove. Most often the only relief from the cold for the infantry was in the form of warming tents set up to the rear of their position.

innersoles and socks. When the Marine halted, the felt innersoles and stockings quickly froze and so did the wearer's feet.

The shoe-pacs were hated, but the sleeping bags that the Marines had carried ashore at Inchon, now had a heavier lining and were much loved and, indeed, were indispensable. Marines found the bags, which could be rolled and tied to the bottom of their haversacks, good for sleeping, for warming feet, and for keeping casualties from freezing to death. There was a problem, though, of Marines standing watch in foxholes being allowed to pull their sleeping bags up to their knees or waists, and then, giving into temptation, slipping further into the bag and falling asleep. A Chinese soldier suddenly upon him killed more than one Marine, caught in his sleeping bag.

In fighting the cold, the Marines learned or relearned certain principles including the importance of keeping moving to generate body heat. The drawback to this was, of course, the sweat-soaked shoe-pacs that invited frostbite. The digging in of a foxhole, which could require six to eight hours of effort, often at the end of a long march, also generated heat, sometimes presenting the paradoxical sight of a Marine, stripped almost to the waist, hacking away at the frozen earth. In last analysis, the imposition of cold weather discipline depended upon the small-unit leadership of lieutenants, sergeants, and corporals. All things considered, they did amazingly well.

The Marines were still using their World War II pack; a well designed but complicated piece of equipment with a haversack, knapsack, a bedding roll, and many straps and buckles. Ordinarily, a Marine in combat carried nothing but his haversack and sleeping bag, and, of course, his rifle belt with its load of canteens, bayonet, first aid packet, ammunition, and possibly a few grenades. Most Marines preferred a pack board whenever they could get one.

A Marine also had to find a place for his daily C-ration when it was issued. It came in a clumsy cardboard box about the size of a shoe box, six cylindrical cans in all, three "heavies" and three "lights," plus an assortment of packets that included a day's supply of toilet paper and a neat little box with four cigarettes. The "heavies" were the meat components, much improved and with a much wider variety of items than the disliked World War II C-ration. The Army's Quartermaster Corps had worked hard on the improvements, basing them on regional favorites. Among the offerings were hamburgers (highly prized), chicken with vegetables, ham and lima beans, meat and beans, and sausage patties (the least favorite). The "lights" included at least one half-sized can of some kind of fruit, easily the best-liked element in the ration and one or more "bread units" which were biscuits of one sort or another, descended from Civil War hardtack, and something that passed for cake. Also to be found were different sorts of candy (disks of chocolate were preferred), salt, pepper, and packets of soluble coffee and cocoa. Most often a Marine took out what he liked or could trade and threw the rest away. What he retained he would fit into various pockets. He wondered why the ration could not be packed in flat cans that he could pocket more easily. His largest problem, though, was heating the meat component. Best method was to heat it in a bucket or GI can of boiling water, but these were seldom available. Cooking fires made with available wood usually did more burning than cooking. Unused mortar increments and bits of C-3 plastic explosive, when they could be found, burned with a quick hot heat. Dirt in a larger can, doused with gasoline, gave an improvised stove. But such open fires did not do well, tending to scorch the meat closest to the can and leaving the interior still frozen. Jeep and truck drivers could wire a can to their engine and when their run was finished, have a hot meal.

C-ration meat components would begin to freeze as soon as their cans were removed from the heat. Drinking coffee from an aluminum mess cup could be a dangerous process, the drinker's lip or tongue freezing to the cup. On the march it was often impossible to heat the meat component. Consequently the bread unit and fruit component were the first to be consumed.

Marines soon learned that keeping a thin coat of oil on their weapons, as taught to them emphatically by their drill instructors at boot camp, was not a good idea in sub-zero temperatures. Even a thin coat of oil tended to congeal and freeze the weapon's action. The word went out to wipe all weapons dry of oil. There was some argument over this. Some Marines thought that an infinitesimally thin coat of oil was best. There were arguments, pro and con, on the advisability of keeping personal weapons in sleeping bags or taking them into warming tents, or leaving them out in the cold.

By and large the weapons of the Marines worked well. A notable exception was the caliber .30 M1 and M2 carbine. Already suspect in World War II, it proved to be a miserable failure in sub-zero weather. Its weak action failed to feed rounds into the chamber, the bolt failed to close, and the piece often failed to fire. The release for its box magazine was a fraction of an inch from the safety. Mittened or cold-stiffened fingers sometimes pressed both, dropping the magazine into the snow. Even when a carbine did fire, the round had no stopping power. Most Marines carrying carbines replaced them as quickly as they could (and most often informally) with the prized M1 "Garand" rifle.

The Browning automatic rifle, M1918A2, continued to be a favorite Marine weapon. It functioned in proportion to the care it was given. Ice tended to form in the buffer group and inside the receiver.

As with all weapons with a recoil mechanism, machine guns, in general, were sluggish in their rate of fire. The old reliable Browning water-cooled M1917A1 fired well as long as there was antifreeze (not always easy to get) in the water jacket. Without liquid, the barrels quickly overheated. The barrels of the M1919A4 light machine gun tended to burn out and there were not enough spares. The 60mm and 81mm mortars fired reliably although there was considerable breakage of base plates and optical sights. It was remarked that the 81mm mortar shells looping across the sky left fiery tails more like rockets.

As to the cold, some units did claim nighttime temperatures of 35 degrees and even 40 degrees below zero. Best-documented temperatures, though, are the records kept by the battalions of the 11th Marines, the artillery regiment, that had to factor in the temperature as an element of gunnery. These battalions routinely recorded temperatures of 20 and 25

Department of Defense Photo (USMC) A5356

As shown in this photo of Marines marching out of Koto-ri on 8 December, each Marine carried what he considered necessary to live and fight, a considerable load of upwards of 60 pounds. Some got along with just their sleeping bag slung below their haversack. More carried a horseshoe-shaped bedding roll that could contain as much as a sleeping bag, a blanket, a poncho, and a shelter half.

below zero. Snow showers were frequent but not much snow accumulated. The winds of 35 and 40 miles per hour tended to blow the rock and frozen earth free of the thin snow. When morning came there would be ice crystals in the air, glinting in the sun like "diamond dust."

Water in five-gallon "Jerry" cans and individual canteens turned into blocks of ice. Some Marines carried a canteen inside their clothing to keep it thawed. Since World War II and the thirst of the Pacific War it had been the Marine Corps habit of having each man carry two canteens. This continued in the Korean War. Some Marine officers and senior noncommissioned officers carried whiskey in their left or "port side" canteen, which they doled out to their subordinates on a most-needed basis. The surgeons also had a carefully controlled supply of two-ounce bottles of medicinal brandy. Those lucky enough to get a bottle might use it to thaw out a C-ration can of fruit and then comment wryly on the luxury of "dining on brandied peaches."

Immersion heaters seldom provided enough warmth to thaw the contents of a water trailer. All valves and piping froze solid. Fires built beneath the trailers were a sometime effective expedient. Some men ate snow. The favorite beverages, when the water for them could be heated, were the soluble coffee and cocoa to be found in the C-ration, or better yet, the more generous allowance in larger rations.

A-rations, the full garrison ration with fresh or frozen meat, fruits, and vegetables, was, of course, unavailable except in an extraordinary set of circumstances such as the celebrated Thanksgiving dinner. B-rations, where canned items replaced the fresh or frozen items, were available but hard to use. Indeed, most of them were wasted, as they required a field kitchen for preparation. Efforts were made in the defensive perimeters to set up consolidated field messes serving hot chow, but this seldom benefited the men actually serving in the frontline. An exception were the flapjacks or pancakes made around the clock by a battalion mess at Hagaru-ri and served to thousands of Marines and soldiers. Artillery batteries sometimes managed to set up their own small messes. Captain Andrew Strohmenger's battery, also at Hagaru-ri, was known for its doughnuts.

Big square cans of ground coffee were a component of the B-ration. These, where space in a jeep trailer could be found, would be kept hoarded until circumstances permitted the boiling up of a batch of real coffee in a can or pail. Oatmeal, also to be found in the B-ration with the cooperation of a friendly mess sergeant, boiled in similar manner and flavored with sugar and powdered milk, was another favorite that riflemen, unfortunately, seldom enjoyed. Canned peanut butter, passed from Marine to Marine and dug out of the can by grimy fingers, was popular and more portable.

As a variant to the C-rations, there were sometimes the larger "five-in-ones" and "ten-in-ones," much

As the march continued south from Koto-ri, the Marines took an increasing number of Chinese prisoners. The Chinese, who had padded uniforms, but little protection for their hands and feet, and no tentage, suffered much more from the cold than did the Americans.

National Archives Photo (USMC) 127-N-A5387

the same in content but with a more varied menu and intended for group consumption by a fire team or squad. For some reason these always seemed to be more available at the higher echelons and seldom at the rifle company level.

Post exchange supplies—cigarettes, candy, writing paper, and such—nominally there to be sold to the Marines, were given to them at no cost in the forward areas. Not many letters were being written, but the candy was a great favorite, particularly the chocolates and hard candies that gave quick energy. Brand-name choices were Tootsie Rolls and Charms.

An enormous advantage that the Marines had over the Chinese was the availability of tentage. The standard tent was the same as used in World War II, four-sided or pyramidal in shape, 16 feet on a side, and with a center pole. A practiced crew could erect one in 15 or 20 minutes even in the cold. Heat was provided by an M1941 stove or space heater the size and shape of a quarter keg of beer. Diesel oil was the preferred fuel, but it thickened in the cold and was frequently—if dangerously—thinned with gasoline to make it flow through the stove's carburetor. The stove stood at the base of the center pole and was good for many things besides heating

Amidst a snowstorm, a 60mm mortar squad rests by the side of the road south of Koto-ri on 8 December. In general, the Marines' winter clothing was cumbersome but effective—except for the shoe-pac. The two Marines in the foreground manage a grin for the cameraman. Note the mittens worn by these Marines.

National Archives Photo (USMC) 127-N-A5359

at least the center portion of the tent; the sidewalls of which were usually, despite the stove, rimed with frost.

General practice was one warming tent per platoon, even rifle platoons. This was possible within the perimeters. Marines were cycled through the tents in relays as frequently as the situation would permit, usually not more than six at a time nor for longer than 20 minutes. They were not a place to sleep. Exact practices varied with location and unit. The warming tents had odd psychological effects. The canvas sidewalls seemed to shut off the war, offering a non-existent protection. Too many Marines could not be clustered together at one time in a tent that might be hit by a mortar shell or machine gun fire. One common practice was to have a communal pot—one company headquarters, as its most prized possession, had a stainless steel pail it used as such, rescued months before from a hospital in Seoul—filled with stew or "slum" constantly simmering on the stove. A Marine, entering the tent, would take his share of the heated slum and then replace it with the contents of one of his C-ration cans. A favorite condiment to make the stew palatable was Tabasco red-pepper sauce, a bottle of which always seemed to materialize. Short sections of wood were often nailed as cross trees to the tent pole as a drying place for sweat-soaked socks and felt shoe-pac liners.

On the march there was some attempt, not very successful, at having warming tents as way stations. Within the perimeters other tents, protected with sandbags, were designated as command posts, usual for regiment and battalion and sometimes, but not often, at the company level. Each perimeter had a field hospital of sorts, using a convenient schoolhouse or some such building. A battalion surgeon might have a cluster of tents, and there was such a thing as a hospital tent, considerably larger than the pyramidal tent. Company-level corpsmen often had a pyramidal tent to use as a sick bay for a few sick, lightly wounded, or exhausted Marines, and as a place to stash their stretchers and medical supplies. Life-saving plasma needed warmth in order to flow. Corpsmen working in the field during a firefight commonly carried morphine Syrettes in their mouths to keep them warm enough for injection.

Elimination of body waste was an unending problem. Within the defensive perimeters at Yudam-ni, Hagaru-ri, Koto-ri, and Chinhung-ni there were certain niceties of expeditionary plumbing available to the headquarters, artillery, and service units if not to the infantry. Packing tubes from mortar and artillery shells provided al fresco urinals. "Four-holers," collapsible plywood "heads" or "shitters," reportedly a Marine Corps invention dating back to the Banana War days, were set up in warming tents. These conveniences, almost never at hand for the rifle units, were not available to anyone on the march out. A much-repeated dark joke involved the problem of finding one's cold-shriveled penis through the many layers of clothing. Urine froze immediately on hitting the cold ground. Defecation was such a difficult procedure that some Marines simply stopped defecating. Later battalion surgeons and company hospital corpsmen would have to contend with impacted colons.

By the time the Marines, after their rehabilitation at Masan, began to move north at the beginning of 1951, some things had gotten better. A small mountain-type gasoline camp stove, about the size and shape of a quart oilcan, was issued on the basis of one stove to every four Marines. It largely solved the task of heating C-rations and at the same time produced boiling water for soluble coffee or cocoa. Inflatable rubber mattresses, to be used as insulation under the much-treasured sleeping bags, also began to appear. They worked best on a canvas cot, but riflemen seldom had the luxury of a cot even when in reserve. Not until the next year, however, would a thermal "Mickey Mouse" boot replace the hated shoe-pac.

Meanwhile, in the United States, the Marine Corps sought a cold-weather training site in California. Big Bear was tried, but serious training did not mix with a ski resort. General Smith, after arriving at Camp Pendleton in May 1951 to be the base commander, took an active personal interest in finding a suitable location. Reconnaissance parties were sent out and by late summer a site was found 450 miles north of Camp Pendleton in the Toiyabe National Forest in the High Sierras. With a valley floor at 6,800 feet, elevations went up to more than 11,000 feet. Weather records promised winter temperatures of 20 below zero and 20 feet of snow. Marines called it "Pickle" Meadow, but it was really Pickel Meadow, named for Frank Pickel, a trapper who had built a cabin there in the 1860s. By fall 1951 all replacement drafts and other units headed for Korea would have a week's in-the-field training at Pickel Meadow.

(Continued from page 25)
Day 1950 was marked in X Corps by the landing at Wonsan of the 15th Infantry, largely schools troops from Fort Benning. The regiment, under Colonel Dennis M. "Dinty" Moore, was to relieve Puller's 1st Marines in and around Wonsan. Almond was not pleased to learn that the 3d Battalion, 15th Infantry, was a "Negro" unit and therefore, in his mind, not completely trustworthy. The 3d Division's third and last regiment, the 7th Infantry, commanded by Colonel John S. Guthrie, came from Fort Devens, Massachusetts, by way of Japan and would disembark at Wonsan on 17 November.

Almond celebrated Armistice Day with an order at midnight calling for an advance to the border, the ROK I Corps on the right, the 7th Infantry Division in the center, and the 1st Marine Division on the left. The Marines were allotted a 40-mile stretch along the Yalu as their ultimate objective. That same day MacArthur indirectly ordered Almond, by way of a personal letter from his G-3, Major General Edwin K. "Pinky" Wright, to do everything he could to assist the Eighth Army in its drive to the Yalu.

Almond had his staff prepare an analysis of the Eighth Army's situation. It credited Walker with having 120,000 troops with which to oppose 100,000 of the enemy and having the advantages of air supremacy and superior artillery support. Almond's study concluded that the severing of the enemy's MSR at Mupyong-ni by X Corps would greatly assist Eighth Army's advance. Almond, in his reply to Wright on 14 November, proposed that he attack north and then west to link-up with the Eighth Army.

Members of the 7th Marines "answer up" at a mail call at Koto-ri on 15 November. A large amount of mail had accumulated, some of it intended for Christmas. Airmail arrived in a prompt five or six days; packages could take five to six weeks or longer. The tall Marine with a letter in his hand is carrying a carbine, a weapon that would prove worthless in the cold weather ahead.
Photo by Cpl L. B. Snyder, National Archives Photo (USMC) 127-N-A4632

As the Chinese Saw It

Peng Dehuai's chief of staff, Xie Fang, at about this time made his own assessment of the situation:

> Our 9th Army Group main forces have successively entered Korea from J'ian and Linjiang to assume eastern front operations. . . . We have over 150,000 men on the eastern front, the enemy over 90,000, giving us a 1.66 advantage over him. We have 250,000 men on the western front, the enemy 130,000, giving us a 1.75 advantage over him. Our forces are superior on the eastern and western fronts.

On 16 November, Xie Fang reported: "Our forces on the eastern front abandoned Hwangch'o [Funchilin] Pass on the 7th. On the 10th. . . the enemy on the eastern front continued advancing northward along three separate routes: From Hwangch'o Pass, P'unsan [Pungsan], and Myongchon . . . still far from our pre-selected killing zones."

Monday, 13 November

Meanwhile, in a division order dated 13 November, Smith directed RCT-1 to take Huksu-ri, RCT-7 to seize Hagaru-ri and on order advance on Yudam-ni, and RCT-5 to protect the MSR from positions at Majon-dong, Chinhung-ni, and Koto-ri, and to be prepared to pass through RCT-7 at Hagaru-ri and advance to Changjin 40 miles to the north.

The road leading north from Koto-ri to Hagaru-ri followed a valley formed by the Changjin River. As Litzenberg's Marines moved on toward Hagaru-ri, 11 miles north of Koto-ri and at the southern tip of the Chosin Reservoir, they could see parties of Chinese in the distance.

On 15 November Rear Admiral Albert K. Morehouse, chief of staff of U.S. Naval Forces, Far East, visited Smith. Smith, feeling he was speaking within the naval family, outlined for Morehouse, to be passed on to Vice Admiral C. Turner Joy, the Commander, Naval Forces, Far East, his concern over what he considered Almond's unrealistic planning and tendency to ignore enemy capabilities. Smith may or may not have shown Morehouse a letter he had just drafted to General Cates.

Alarmed at the prospect of attacking simultaneously in two different directions, Smith had stepped out of the chain-of-command to write a personal letter to the Commandant of the Marine Corps. In it he said:

> Someone in high authority will have to make up his mind as to what is our goal. My mission is still to advance to the border. The Eighth Army, 80 miles to the southwest, will not attack until the 20th. Manifestly, we should not push on without regard to the Eighth Army. We would simply get further out on a limb. If the Eighth Army push does not go, then the decision will have to be made as to what to do next. I believe a winter campaign in the mountains of North Korea is too much to ask of the American soldier or marine, and I doubt the feasibility of supplying troops in this area during the winter or providing for the evacuation of sick and wounded.

Photo by Cpl Peter W. McDonald, National Archives Photo (USMC) 127-N-A4875

Marines found the 75mm recoilless rifle, shown here in action near Koto-ri in mid-November, of increasing use in the mountains. It gave them a direct-fire weapon of great accuracy and lethality. But the shells were heavy and difficult to lug up into the hills in any great number.

In conclusion, Smith underscored his concern over "the prospects of stringing out a Marine division along a single mountain road for 120 air miles from Hamhung to the border."

Asked years later to comment on this extraordinary action by Smith, Almond said tartly: "My general comment is that General Smith, ever since the beginning of the Inchon landing and the preparation phase, was overly cautious of executing any order that he ever received."

In 1952, General Shepherd, by then Commandant, would report to the Secretary of the Navy: "By orders of higher authority the division was placed in a situation, which, when the Chinese struck in force on 28 November 1950, resulted in the division being in effect deployed in column for a distance of 35 miles within enemy territory. . . . The wide separation of elements of the Tenth Corps of which the First Marine Division was a part, and the gap existing between the Tenth Corps and the Eighth Army had permitted the Chinese to flow around the First Marine Division preparatory to an all-out attack."

MacArthur, responding to Almond's 15 November proposal, asked Almond for an alternate plan giving priority to taking off the pressure confronting the Eighth Army. Accordingly, Almond now visualized an attack by the 1st Marine Division on the Hagaru-ri—Mupyong-ni axis with a regimental combat team from the 7th Division protecting the division's right flank by taking Changjin. This became the operative plan. Almond recognized that extreme minimum temperatures of from 30 to 40 degrees below zero would severely restrict both friendly and enemy operations.

Photo by TSgt J. W. Helms, Jr., Department of Defense Photo (USMC) A155763

A Marine squad trudges through the snow-encrusted streets of Hagaru-ri. The 7th Marines occupied the town on 15 November. By then the weather had turned cold. Weather records indicated that Hagaru-ri could be the coldest spot in North Korea.

7th Marines Reach Hagaru-ri

While the commanders exchanged proposals and plans, the 7th Marines occupied Hagaru-ri on 15 November. The nighttime temperature had dropped to four degrees below zero. Hagaru-ri was a medium-sized town, fairly well flattened by bombing. Just north of Hagaru-ri in the hamlet of Sasu-ri there was a sawmill and a great deal of fresh-cut lumber. Once tents began to spring up, the town reminded at least one Marine officer of an Alaskan gold camp with its mud-and-snow streets, its tents, and rough construction with raw lumber. General Craig visited Hagaru-ri and recommended it to Smith as a forward base.

By then RCT-5 had its 2d Battalion at Koto-ri, its 3d Battalion at Chinhung-ni—along with much of the remainder of the division—and its 1st Battalion at Majon-dong. As Murray, the regimental commander, remembered:

> We'd been highly successful in the south, and we had a lot of this carry over as we went north. There wasn't anybody any better than we were, that was the general feeling in the regiment. . . the hills seemed to be a lot steeper than they were in the south. . . . And in some cases, on the road between, I guess it was just below Hagaru-ri a ways, there was a power plant built right into the side of the mountain, and the road ran over a part of this thing. Very easy to blow it up, which was done, done twice as a matter of fact by the Chinese later on.

Smith again visited the Chosin plateau on 16 November, this time driving up in a heated station wagon. At Chinhung-ni he met, by coincidence rather than design, Major General Field Harris, 55, the commanding general of the 1st Marine Aircraft Wing. Harris had flown as far as Chinhung-ni in a helicopter and had planned to go the rest of the way by open jeep. Smith offered him a ride in his station wagon. They drove comfortably to Hagaru-ri with Smith in a rare burst of jocularity promising

A tent camp sprang up at Hagaru-ri. On 18 November there was time for a brief awards ceremony. Col Homer Litzenberg, right, and LtCol Raymond G. Davis, left, Commanding Officer, 1st Battalion, congratulates Cpl Earle R. Seifert, who received a Bronze Star, and SSgt Earle E. Payne, a Navy Commendation Ribbon, for earlier heroism.

Photo by Cpl L. B. Snyder, National Archives Photo (USMC) 127-N- A4641

National Archives Photo (USN) 80-G-421479

VAdm C. Turner Joy, left, is greeted at Wonsan on 19 October by MajGen Field Harris, commanding general of the 1st Marine Aircraft Wing. Joy, as Commander, Naval Forces, Far East, was MacArthur's naval component commander. No direct command line linked Harris and the wing to O. P. Smith and the 1st Marine Division.

move up the east side of the reservoir and seize Sinhung-ni, about seven miles northeast of Hagaru-ri. (Sinhung-ni, just east of Chosin Reservoir should not be confused with Sinhung in Sinhung Valley previously visited by the 5th Marines.)

Murray had been told to nominate a battalion commander for return to the United States. He picked George Newton, commander of the 1st Battalion. Murray said of Newton: "He was a very competent battalion commander, but he was, I felt, almost killing himself trying to be a good battalion commander. He seemed to stay awake most of the time."

"George left [on 17 November] before we went all the way up," said Murray. "Anyway, George Newton was relieved by a pretty good leader [Lieutenant Colonel John W. Stevens II]. But I did have good battalion commanders. We had an excellent staff. The main thing, as I say, is that we had been

Harris a station wagon of his own in exchange for continued close air support.

Almond had asked Field Harris to reconnoiter Hagaru-ri for a site for an airstrip long enough to handle two-engine transports. Smith and Harris walked the ground and found a stretch just south of the town that seemed suitable. "There is plenty of room, but the soil consists of a thick, black loam," Smith entered in his log. "If the ground freezes it will probably be all right for a strip."

Regiments Get New Orders

On 17 November Smith modified his orders to his regimental combat teams. RCT-7 was to protect the left flank of the division between Hagaru-ni and Yudam-ni. RCT-5 was to pass a battalion through RCT-7 at Hagaru-ri and

An M-4A3 Sherman tank from the 1st Tank Battalion travels along a well-graded but narrow road coming up Funchilin Pass on 19 November. Tanks gave the Marines enormous firepower and were useful in crushing enemy roadblocks, but weather and terrain tended to keep them road-bound.

Photo by TSgt J. W. Helms, Jr., National Archives Photo (USMC) 127-N-A5343

National Archives Photo (USN) 80-G-423190

MajGen O. P. Smith and RAdm James H. Doyle, shown here at the time of the Inchon landing, were long-time friends. Both were highly experienced in amphibious warfare. They formed an effective and compatible partnership not only at Inchon, but also in both the Wonsan landing and the ultimate evacuation from Hungnam.

successful in the south, and all that was needed was to keep this going." Stevens was a known quantity. He had been the executive officer of the 2d Battalion, 5th Marines, from Pusan on through Inchon and Seoul.

That evening Smith dined on board the amphibious command ship *Mount McKinley* (AGC 7) with Rear Admiral James H. "Jimmy" Doyle. Describing Doyle as "a typical Irishman," Colonel Bowser, Smith's G-3, said: "He is a real fighter when it comes to the clutches. A fun guy to know—always a laugh or a joke." Smith and Doyle, alone in the admiral's cabin, in Smith's words, "let our hair down."

Vice Admiral Arthur D. Struble, commander of the Seventh Fleet, had been superimposed over Doyle's Task Force 90 during the Inchon and Wonsan landings. Doyle disliked Struble and, dubious of his competence, was determined to keep the Seventh Fleet out of direct control of future amphibious operations. After the Wonsan landing, Doyle had complained to his old friend Vice Admiral Joy, commander of Naval Forces, Far East, that he could not and would not come under Struble again. He was successful in his arguments. As commander, Task Force 90, at Hungnam he would report direct to Admiral Joy.

By now engineers had improved the MSR to a point where armor could be sent forward to join Litzenberg. A tank platoon reached Hagaru-ri on 18 November. That same day Smith visited Puller at his command post just west of Chigyong. Smith noted that there was snow on the mountains but that the road was still open. He was resisting an order from Almond to send a battalion to Huksu-ri, about 20 miles to the northwest, to occupy a blocking position. "There is no truck road to take," said Smith in his log. "I do not intend to put Puller out on a limb where he cannot be supplied. Also I would like to close him up behind the regiments moving toward the Chosin Reservoir. The 26th ROK Regiment is attacking toward Huksu-ri. Possibly this will relieve me of concern regarding that place."

Construction of the airstrip at Hagaru-ri began. Smith asked for X Corps engineers, but could get none. The job was given to Lieutenant Colonel John Partridge's 1st Engineer Battalion. Wind-blown Hagaru-ri was at an elevation of about 4,000 feet. For that altitude the engineer manuals prescribed a minimum runway of 3,900 feet for C-47 transport operations. The engineers crossed their fingers and hoped that a strip as short as 3,000 feet might do. Once started, construction of the airstrip proceeded 24 hours a day, with work at night under floodlights.

Marine observation squadron, VMO-6, although part of the 1st Marine Aircraft Wing, was under Smith's operational control. Smith regarded the squadron as his own private air force. On the 19th, he visited the squadron's commander, Major Vincent J. Gottschalk, at Yonpo airfield to discuss the problems of operating helicopters and light aircraft in the cold at high altitudes. Gottschalk, 31, promised to provide solutions. He had come into the Corps in 1941 after graduating from the University of Michigan. For much of World War II he had served as Marine detachment commander in the light aircraft carrier *Langley* (CVL 27)—after the war came two years of flight training.

Early in November, Admiral Joy had asked Smith if he could use the Royal Marines' 41 Independent Commando—14 officers and 221 enlisted men, commanded by Lieutenant Colonel Douglas B. Drysdale. Smith replied he would be glad to get these fine troops, foreseeing 41 Commando operating with the division Reconnaissance Company in screening the flanks of the Marine advance. The British Marines arrived at Hungnam on 20 November, the same day that Almond passed on instructions from higher headquarters that "damage, destruction or disruption of service of power plants will be avoided." In the larger scheme of things, the intention was to leave the hydroelectric generators intact. Marines would wonder why.

On 21 November the division's southern boundary was adjusted to give the responsibility for Huksu-ri to the 3d Infantry Division. Puller's regiment was now available to fill in behind Murray and Litzenberg.

Photo by Sgt John Babyak, National Archives Photo (USMC) 127-N-A4683

The well-meaning, but bumbling Secretary of the Navy Francis P. Matthews, greeted here by BGen Craig, visited the division's rear area at Hamhung on 22 November. At the division hospital he was surprised to learn that Navy personnel met the Marine Corps' medical needs. Another politician accompanied him, Senator Claude Pepper of Florida.

Secretary of the Navy Visits

Wednesday morning, 22 November, found O. P. Smith and Field Harris on the Yonpo airfield awaiting the arrival of the Secretary of the Navy, Francis P. Matthews. Behind his back, Matthews was known as "Rowboat" because of his lack of knowledge of naval matters. Accompanying the secretary was Admiral Joy and Senator Claude Pepper of Florida. Arriving at the airfield at the same time was President Syngman Rhee. Matthews had wanted to call on Rhee in Seoul but could not get clearance from the Secretary of State, Dean Acheson, to do so. This left Smith and Harris with the ticklish problem of keeping the two high-level parties apart. They whisked Matthews away from the field before he could learn of Rhee's presence, taking him to the division hospital. There were very few wounded Marines to visit, but Matthews found the Chinese and North Korean prisoner of war patients of great interest. He seemed to have difficulty understanding why Navy personnel were running a Marine hospital. It was a picture-taking opportunity for the secretary followed by another picture-taking opportunity at the division cemetery. Next the division staff gave the visitors a briefing followed by

LtCol Douglas B. Drysdale, RM, and his 41 Independent Commando, Royal Marines, were billeted briefly with the 1st Engineer Battalion in Hamhung before moving up to the Chosin Reservoir. Drysdale's command, largely made up of volunteers, had assembled in Japan where it was re-equipped with American infantry weapons.

Photo by TSgt J. W. Helms, Jr., National Archives Photo (USMC) 127-N-A5322

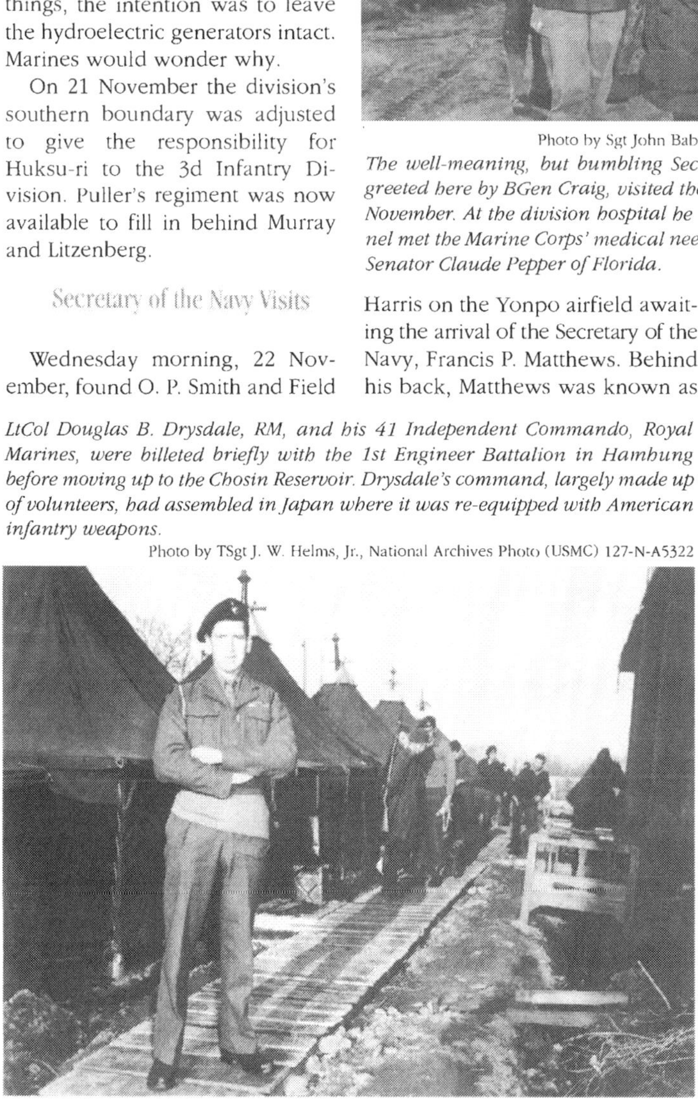

Photo by Sgt Peter Ruplenas, National Archives Photo (USA) 111-SC363307

It was a proud moment for MajGen Almond, center, when, on 21 November, soldiers of the 7th Infantry Division reached the Yalu River. Savoring the moment with him, from left to right: BGen Homer Kiefer, division artillery commander; BGen Henry I. Hodes, assistant division commander; MajGen David G. Barr, division commander; and Col Herbert B. Powell, Commanding Officer, RCT-17.

lunch in the commanding general's mess. The cooks had embellished the standard ration with biscuits and cookies. Secretary Matthews and Senator Pepper seemed to enjoy these immensely. Secretary Matthews was then escorted back to Yonpo airfield. Smith again managed to keep him unaware of President Rhee who was departing at the same time. Senator Pepper stayed on for the rest of the day. He wanted to visit with some Marines from Florida. Smith found 15 of these. He had not been able to find any Marines from Nebraska for Secretary Matthews.

Thanksgiving, 23 November

Thanksgiving fell on Thursday, 23 November. The holiday menu, accomplished by strenuous effort on the part of many hands, included shrimp cocktail, stuffed olives, roast young tom turkey with cranberry sauce, candied sweet potatoes, fruit salad, fruit cake, mincemeat pie, and coffee. Even the Marine infantry units got at least the turkey.

Admiral Doyle sent in a cooked turkey for General Smith's mess, but Smith himself had been invited to dinner by Almond. As Smith said in his log: "The dinner was complete with cocktails served from a cocktail bar, tablecloths, napkins, Japanese chinaware, regular silverware, place cards, etc. Admiral Struble and Generals Biederlinden (G-1 of GHQ), Harris, Barr, and Ruffner were also present."

Two days before Thanksgiving, elements of the 7th Division's 17th Regiment had reached the Yalu without encountering a single Chinese soldier. Years later General Almond still savored that moment of triumph:

> And on the 21st of November the leading battalion of the 17th Infantry reached the Yalu River and I was present when they did so. . . . I accompanied General Barr, the division commander; General Hodes, the assistant division commander; and General Kieffer, the artilleryman; with the regimental commander, Colonel Powell. We all walked behind the lead company down the road to the river bank. This was the first element of the American forces to reach the Korean-Manchurian border, although earlier elements of the 6th ROK Division with I American Corps on the west flank, Eighth Army front, attempted to get to the river but did not succeed in remaining there.

Almond and his commanders paused on the banks of the Yalu for a ritual urination into the waters of the river. Meanwhile, Colonel Charles E. Beauchamp's 32d Infantry was advancing to the northwest of Powell's 17th Infantry with orders to reach Singalpajin, originally a Marine Corps objective, on the Yalu. A 34-man patrol under Second Lieutenant Robert C. Kingston (a future four-star general) was sent out from the 3d Battalion, 32d Infantry. The patrol reached Samsu, 23 miles south of the Yalu, where it held on for three days, and then, reinforced by tanks, artillery, engineers, and more infantry, plunged forward, still commanded by the 22-year-old second lieutenant. Now designated "Task Force Kingston," it arrived at Singalpajin on 28 November, fought a house-to-house fight with North Koreans, and then took its turn at urinating in the Yalu. The second and last American unit to reach the Chinese border, Task Force Kingston, for all of its adventures, suffered only one casualty: a soldier reportedly killed by a Siberian tiger.

While soldiers and Marines were eating their Thanksgiving turkey, Smith again modified his orders for the 1st Marine Division's advance. RCT-7 was to move on to

Photo by TSgt J. W. Helms, Department of Defense Photo (USMC) A5289

In the 1st Marine Division rear at Hamhung, where the weather was still benign, members of Maj Robert L. Schreier' 1st Signal Battalion queue up for Thanksgiving dinner. This battalion, responsible for both the wire and radio communications of the division, also included ANGLICO—the Air and Naval Gunfire Liaison Company that provided forward observer teams to Army and ROK elements of X Corps.

Yudam-ni. RCT-5 was to continue up the eastern side of the reservoir. RCT-1 was to protect the MSR from positions at Hagaru-ri, Koto-ri, and Chinhung-ni. As Smith said in his log:

> I did not want to push Murray too far or get Litzenberg out on a limb at Yudam-ni until I could close up Puller in rear of them. . . . I had hoped there might be some change in the orders on the conservative side. This change did not materialize and I had to direct Litzenberg to move on to Yudam-ni.

Most of the 7th Marines had their Thanksgiving dinner at Hagaru-ri. The 2d Battalion, 7th Marines, had set up its field mess in the shadow of what would come to be called "East Hill." Private First Class Alfred P. Bradshaw, a reservist who had recently joined Captain Hull's Dog Company, had lost his mess gear. The mess kits consisted of two flat aluminum pans clamped together, not much changed in pattern since the Civil War. Marines in rifle companies seldom had need for mess gear; they subsisted almost entirely on C-rations, thankfully much improved since World War II. Bradshaw, standing in the chow line, had sought to improvise a plate out of a piece of cardboard. Hull saw Bradshaw's plight and gave him one of his pans. Bradshaw would remember that.

The road from Hagaru-ri to Yudam-ni climbed up through Toktong Pass, four miles to the northwest and about 4,000 feet in elevation, and then descended into a narrow valley before reaching Yudam-ni. Smith personally gave Litzenberg orders to drop off a company at Toktong Pass.

On the day following Thanksgiving, 24 November, MacArthur came to Korea to see the jump-off of the Eighth Army on the offensive that was to end the war. He announced to the press that the war would be won in two weeks and that the Eighth Army would spend Christmas in Japan. To complete Walker's victory, MacArthur ordered Almond to execute the already planned attack to the west so as to squeeze the Chinese

Elements of both the 5th and 7th Marines spent Thanksgiving within the perimeter of the burgeoning combat base at Hagaru-ri. Every effort was made to reach every Marine in the division with a traditional holiday dinner. Here a 5th Marines cook ladles pumpkin pie mix into a piecrust spread out in a square pan.

National Archives Photo (USMC) 127-N-A4726

Department of Defense Photo (USMC) A6791

Thanksgiving was a last lull before the Chinese storm broke. At Hagaru-ri, Reverend Lee In Sup, a Presbyterian pastor, and his wife joined the 5th Marines for Thanksgiving services. Lee thanked LtCol Murray, commanding officer of the 5th Marines, for the liberation "of our country and our church." Beaming broadly in the background is the regimental chaplain, LtCdr Orlando Ingvoldstad, Jr.

between the Eighth Army and the still-independent X Corps. Lieutenant Colonel John H. Chiles, USA, Almond's G-3, had carried the final draft of X Corps operations order to Tokyo on Thanksgiving Day. MacArthur approved the plan on Friday.

On Saturday morning, 25 November, O. P. Smith attended a briefing at X Corps headquarters outlining X Corps Operation Order Number 7. He learned that his division was to be the northern arm of a giant pincer. The other arm of the pincer would be the Eighth Army. He was to sever the enemy's lines of communication at Mupyong-ni and then advance to the Yalu. He was to launch his attack on Monday, 27 November. Concurrently, the 7th Division would continue its advance northward to the Yalu. Almost 100 miles separated the two divisions. Strength returns for that day showed the 1st Marine Division as having 25,323 Americans with 110 South Koreans attached, but of that number only about 15,000 were up at the reservoir. Indeed, some units of the division were as far to the rear as Japan. A goodly number of hospitalized Marine patients were also carried in the total. The 7th Division strength on the same day was 16,001 men of whom 6,794 were South Korean KATUSA soldiers.

Smith estimated the road distance from Yudam-ni west to Mupyong-ni, over another mountain pass and then through a narrow valley, as being 55 miles. The division was then to advance northward to the Yalu. Almond's three columns—the 1st Marine Division, the 7th Infantry Division, and, nominally under his control, the ROK I Corps—were diverging like the ribs of an opened fan. The 7th Infantry Division was to complete its advance to the Yalu. The ROK corps was to advance to the Chinese border from the Hapsu and Chongjin areas. To the rear the newly arrived 3d Infantry Division, under General Soule, was given a multiplicity of missions: gain contact with the right flank of the Eighth Army; protect the left flank of X Corps; support the 1st Marine Division on order; protect the harbor and airfield at Wonsan; and destroy guerrillas in its zone of action. The 3d Division was also to have had the task of

By the third week in November a tent camp, mostly for combat service units had sprung up at Hagaru-ri. One observer said that the badly battered town reminded him of an Alaska gold-rush camp. In the foreground a bit of the narrow-gauge railroad track that once served Hagaru-ri can be seen.

Photo by Sgt Frank C. Kerr, National Archives Photo (USMC) 127-N-A4971

National Archives Photo (USA) 111-SC352938

Gen Douglas MacArthur came to Korea on 24 November to see the jump-off of the Eighth Army in the offensive that was to end the war. Two days later the Eighth Army was in full retreat. LtGen Walton H. "Johnnie" Walker, commanding general of the Eighth Army, seated behind MacArthur, would die in a traffic accident one month later.

Smith's rough plan was to have the 5th Marines pass through the 7th Marines at Yudam-ni and then attack to the west. The 1st Marines, in reserve, was to occupy positions along the MSR at Chinhung-ni, Koto-ri, and Hagaru-ri. Supporting this plan, Almond decided that a regimental-sized force from Barr's 7th Division should relieve Murray's 5th Marines on the east side of the reservoir so that the 5th Marines could join the 7th Marines at Yudam-ni. He ordered Barr to send a regimental combat team for this purpose by 27 November.

Barr, acting on local intelligence that the Chinese in massive numbers had crossed the Yalu at Linchiang and were moving into the gap between his division and defending the area south of Hagaru-ri, but, with its other missions, the best it could promise to do was take over the security of the MSR from Sudong back to Hamhung. It bothered Smith that the 3d Infantry Division had not yet closed behind him and that he would have to leave Puller's 1st Marines strung out along the MSR to keep it open from Hagaru-ri south to Chinhung-ni.

Advances on Both Sides of the Reservoir

Davis with his 1st Battalion, 7th Marines, had led off the advance to Yudam-ni on Thanksgiving Day. He ran into a defense of Toktong Pass by an estimated 150-200 Chinese, but scattered it with the aid of air and artillery. The battalion paused to celebrate Thanksgiving a day late, and then moved on into Yudam-ni on the 25th against negligible resistance. The 3d Battalion, 7th Marines, and Litzenberg's regimental headquarters followed Davis' battalion into the forlorn village.

LtCol Don C. Faith, Jr., USA, right, commanding officer of the 1st Battalion, 32d Infantry, 7th Division, pictured with the regiment's commanding officer, Col Allan D. MacLean, in Japan in the spring of 1950. Except for limited experience during the battle for Seoul, Faith had not commanded a unit in combat prior to the Chosin Reservoir.

Courtesy of Col Erwin B. Bigger, USA (Ret)

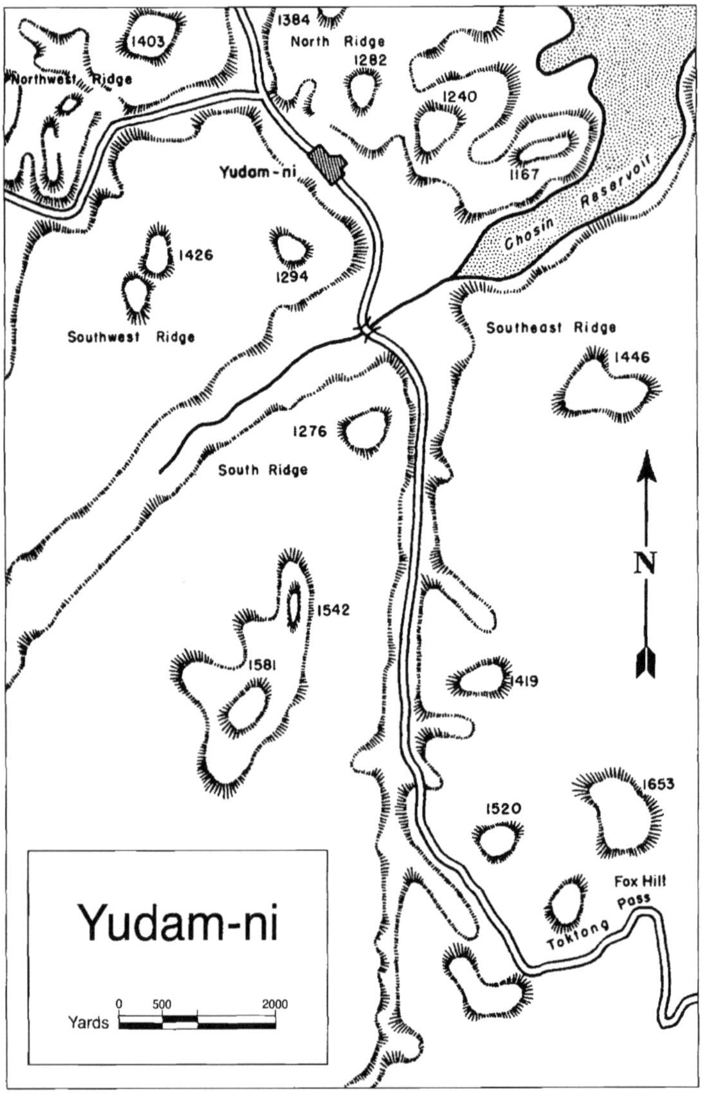

the Marines, had already begun pulling together his scattered battalions.

RCT-31, as assembled by Barr and commanded by Colonel Allan D. MacLean, consisted of the 31st Infantry's Headquarters and Service Company, the regiment's 2d and 3d Battalions, the 31st Tank Company, the 57th Field Artillery Battalion, Battery D of the self-propelled 15th Antiaircraft Artillery Automatic Weapons Battalion, and the 1st Battalion, 32d Infantry, commanded by Lieutenant Colonel Don C. Faith, Jr.

Don Faith, 32, six-feet-tall, handsome, and charismatic, was something of an Army golden boy. The son of an Army brigadier general, he had enlisted in 1941 and won his commission as a second lieutenant the following year. For three years of World War II, he served first in the 82d Airborne Division and then in the XVIII Airborne Corps as an aide to Major General Matthew B. Ridgway with whom he landed at Sicily and jumped into Normandy and Holland. Faith had worked for Barr in China. He had commanded the 1st Battalion, 32d Infantry, for more than a year. In this new war he had been recommended for a Distinguished Service Cross for his performance between Inchon and Seoul.

Barr chose to pull Faith's battalion from the 32d Infantry and assign it to RCT-31 because it, in bivouac northeast of Hamhung, was the Army battalion closest to the reservoir. Faith on 24 November had a strength of 715 Americans and about 300 South Koreans.

Most of Faith's officers were well trained and combat experienced. Some had served in Europe during World War II, some in the Pacific. There was also a layer of battle-hardened senior noncommissioned officers. The mix of Americans and South Koreans in the rank-and-file, however, was a problem, both in language and lack of training. The battalion, in its equipment and preparation for a winter campaign, was about on a par with the Marine battalions, in some ways better and in some ways not as good. "They were short of chains for their trucks. The only tentage they had were tent flies for their kitchens," said one observer. During the previous winter the 31st Infantry, stationed at Camp Crawford, Hokkaido, had received cold-weather training. Most of the men were issued Army winter parkas, shorter and less clumsy than the Navy parkas worn by the Marines. They had sweaters and pile liners of various sorts and shoe-pacs which were really rubber-and-leather hunter's boots. Believed by the troops to have been provided by L. L. Bean, these

National Archives Photo (USA) 111-SC352537

Ill-fated RCT-31, under Col Allan D. MacLean, USA, right, followed RCT-17 ashore at Iwon. Here, on 12 November with the weather still mild, MajGen Almond gives MacLean some words of advice after presenting him with a Silver Star. MajGen Barr, MacLean's division commander, stands stolidly in the center. Two weeks later MacLean would be dead.

boots had been suitable in World War II in the wet cold of northern France and Germany. But, as the Marines were also learning, they were worse than no good in subzero temperatures. Faith himself did not like the shoe-pacs and wore galoshes over his leather combat boots. So did many other soldiers and Marines if they could get them.

On Saturday morning, 25 November, Faith and his battalion, the lead element of RCT-31, started up the icy road to the reservoir. At Hagaru-ri they took the right-hand fork in the road. Some miles up the eastern side of the reservoir Faith met Murray, the commander of the 5th Marines. Murray outlined for him the disposition of his three battalions, all of which were now east of the reservoir. Taplett's 3d Battalion, in the lead, had a good defensive position about four road miles north of the Pungnyuri-gang inlet. Earlier that day, a patrol from Taplett's battalion had almost reached the northern end of the reservoir before brushing up against a small party of Chinese. Murray designated an assembly area for Faith's battalion near the village of Twiggae. Faith set up his command post in a hut on a lower slope of Hill 1221.

With the relief of RCT-5 by Faith's battalion, Marine operations east of the reservoir would end. There was no sign of large-scale enemy activity. The soldiers were to stay under the operational control of the 1st Marine Division until the arrival of Colonel MacLean, the commanding officer, 31st Infantry. Faith's command relationship to the 1st Marine Division was not clear. He asked Murray for instructions. Murray, who did not consider Faith to be under his command, said that he had none, but he did caution Faith not to move farther north without orders from the 7th Division. Once Murray departed, the only radio link between Faith and the 1st Marine Division would be that provided by his attached tactical air control party, led by Marine Captain Edward P. Stamford. He and his four-man team had been with the battalion since Seoul.

Just before noon on the 26th, Brigadier General Henry I. Hodes, the 7th's assistant division commander, visited Faith at Hill 1221. Hodes, 51 years old, a West Pointer who had commanded the 112th Infantry in Europe in World War II, and a future four-star general, told Faith that MacLean and the rest of the 31st RCT would soon be arriving.

Smith Visits Yudam-ni

On Sunday morning, 26 November, Smith visited Yudam-ni. During the night he had been informed that the ROK II Corps, on the right flank of the Eighth Army, had been thrown back in the vicinity of Tokchon, about 70 air miles southwest of Yudam-ni. But as yet Smith had no notion of the extent of the disaster that had befallen the Eighth Army. Both Walker's G-2 and GHQ in Tokyo had badly underestimated the

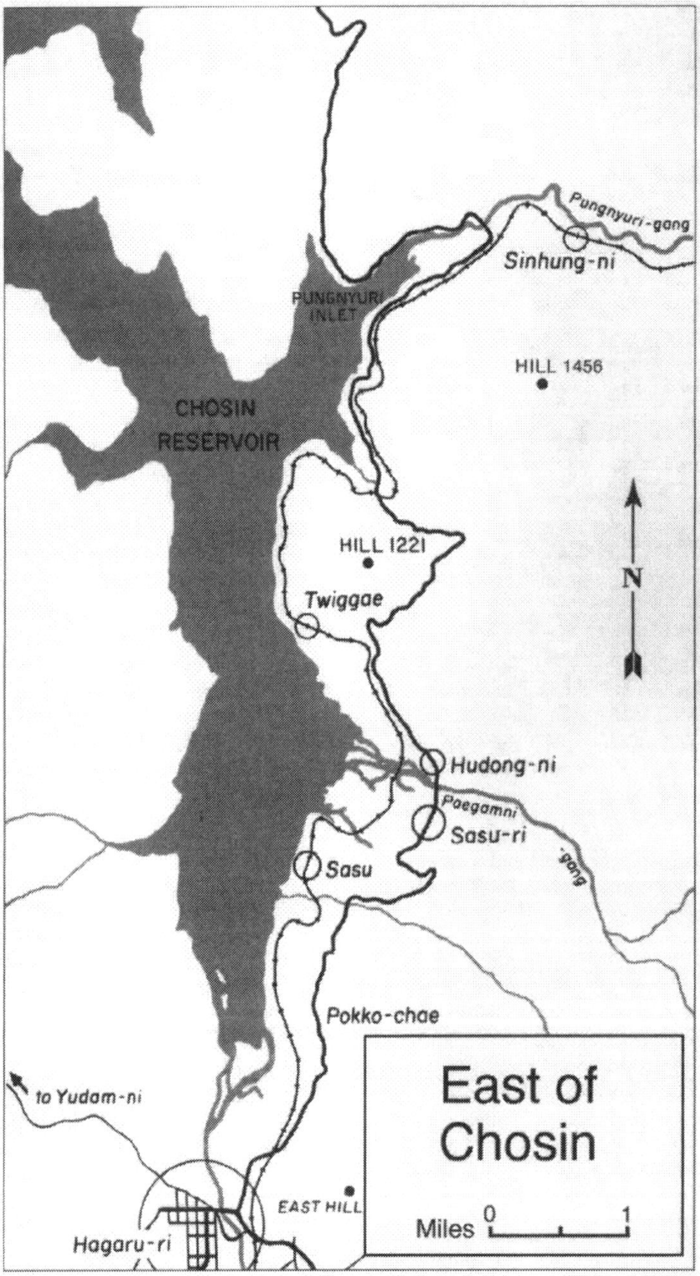

strength of the Chinese. One day into the offensive that MacArthur had blithely informed the press would end the war, the Chinese *Thirteenth Army Group* with 18 divisions counterattacked Walker.

From his helicopter on the way to Yudam-ni Smith could see no signs of enemy activity. As he entered in his log:

> I landed at what I thought was the CP of the 7th, but it proved to be the CP of the 1st Battalion, 7th. I had a visit with LtCol Davis, the Commanding Officer, and got directions from him as to the location of the CP of the 7th, which was about 5000 yards south, up the road to Hagaru-ri. In making the landing at the regimental CP I discovered some of the limitations of helicopters. We first attempted to land on a gentle slope near the CP. As the pilot put his wheels down we slipped backwards on the ice and snow. After 4 or 5 tries we went down to the floor of the valley to land. The elevation here was about 4000 feet. At this altitude the helicopter does not have much hovering capability. There was no air stirring in the bottom of the valley and for the last 10 feet we simply dropped. We hit with quite a bump but no damage was done. Had there been a breeze it might have assisted us in hovering. Litzenberg's role now is to hold the Yudam-ni area while Murray passes through him to continue the advance to the westward. Litzenberg indicated he would like to keep on going.

Yudam-ni lay in the center of a broad valley surrounded by five great ridgelines. Moving counterclockwise from the north, the ridges were given the prosaic but useful designations North, Northwest, Southwest, South, and Southeast. The 7th Marines held a perimeter that commanded four of the five ridges—all but the Northwest Ridge. Yudam-ni itself was a miserable collection of mud-and-thatch houses, battered by air attacks and now abandoned by their owners. The road that was the lifeline of the 1st Marine Division forked at Yudam-ni. One fork continued to the north. The other opened to the west, going as far as Mupyong-ni, before turning north and continuing to Kanggye.

On 26 November the 7th Marines reported the capture of three Chinese soldiers from the *60th CCF Division* and learned from them that the *58th, 59th,* and *60th CCF Divisions,* making up the *20th CCF Army,* were in the vicinity of Yudam-ni.

1st Marines Button Up Division Rear

RCT-1 had to wait several days for rail transport to take them the 70 miles north from Wonsan to Chigyong. The regiment's 1st Battalion relieved the 3d Battalion, 5th Marines, at Chinhung-ni on Thanksgiving. Two days later the regiment's 2d Battalion, along with Puller's regimental headquarters, took over Koto-ri from the 2d Battalion, 5th Marines. Smith now had his regiments fairly close together, but further movement was hindered by a shortage of motor transport.

Two-thirds of the 3d Battalion, 1st Marines, arrived at Hagaru-ri during the early evening of Monday, 26 November. The battalion commander, Lieutenant Colonel Thomas L. Ridge, 35 and University of Illinois 1938, had been a naval attaché in Brazil for much of World War II but had reached the Pacific as an intelligence officer in time for Iwo Jima and Okinawa, where he was twice wounded. The motor march to Hagaru-ri was uneventful except for snarls in traffic. Because of the shortage of trucks, Captain Carl L. Sitter's George Company, reinforced with a provisional platoon from Weapons Company, had to be left behind at Chigyong.

Relief of Lieutenant Colonel Randolph S. D. Lockwood's 2d Battalion, 7th Marines, had to wait until morning. Lockwood, 37, U.S. Naval Academy 1936 and Harvard 1940, had just taken over the battalion from Major "Buzz" Sawyer on 9 November. Lockwood had spent most of World War II as a staff officer in Hawaii. The combat-experienced Sawyer stayed on as battalion executive officer.

On 25 November, the 2d Battalion, 1st Marines, along with Col Puller's regimental headquarters, relieved the 2d Battalion, 5th Marines, at Koto-ri. Next day, this heavy machine gun squad, with its water-cooled Browning M1917A1, follows behind two well-deployed rifle platoons making a reconnaissance in force toward the first range of hills.

Photo by Cpl W. T. Wolfe, National Archives Photo (USMC) 127-N-A4866

The new-arrivals at Hagaru-ri watched the engineers hack away at the frozen earth in their effort to build an airstrip capable of handling Air Force C-47 and Marine R-4D transports. The 1st Medical Battalion under Commander Howard A. Johnson set up a clearing station close to the strip for the expected flow of casualties. Extra surgical teams were flown into Hagaru-ri. The hospital ship *Consolation* (AH 15) moved up to Hungnam from Wonsan. The 1st Marine Division 400-bed hospital at Hungnam had an annex of 150 more beds at Hamhung.

Smith informed Fleet Marine Force, Pacific, and Headquarters Marine Corps, that, unless he received word to the contrary, he was sending his assistant division commander, Brigadier General Craig, home on emergency leave. Craig had received the bad news that his father had suffered a cerebral thrombosis and that the prognosis was unfavorable. Craig left for the States on Monday morning, 27 November.

The Chinese

The Marines were gradually learning about the new enemy. The term used for them by the U.S. and other English-speaking forces was "CCF" or "Chinese Communist Forces." Marines would learn that a CCF division, with its three infantry regiments and an artillery battalion (more theoretical than real in 1950), numbered about 8,000 men. A CCF regiment would average about 2,200 men, organized into three infantry battalions, sometimes with an artillery battery, more often with a mortar company, and several meager support companies. In the forward areas the Chinese had little or no motor transport. Things were pulled in carts by man or beast or carried on

Marine Corps Historical Center Photo Collection

Contrary to popular belief, the Chinese did not attack in "human waves," but in compact combat groups of 50 to 100 men. Here one such group makes its way up a snow-clad hill. The 1st Marine Division came into contact with the major portion of the Chinese Ninth Army Group, *which, with 12 divisions, totaled about 150,000 men.*

the backs of men, either Chinese soldiers or impressed Korean porters. The CCF infantry battalion, on paper at least, looked much like the Marines' own battalions: three rifle companies and a machine gun or heavy weapons company. The rifle companies similarly had three rifle platoons and a 60mm mortar section or platoon. The individual Chinese soldier was physically tough, uncomplaining, and used to long marches with few if any creature comforts. Politically he had been thoroughly indoctrinated, but once taken prisoner that indoctrination would tend to crack.

Collectively, his armament was a mixed bag of weapons gained from the surrender of the Japanese, the collapse of the Chinese Nationalist government—and its mixture of American, British, German, Czech, and other weapons—and the more recent issue of Russian weapons by the Soviet Union. But the Chinese army, at least in this stage of the war, was never equipped as uniformly or as well as the North Korean army had been. For the most part, the Chinese soldier wore a two-piece padded uniform with a cap to match, fairly adequate of themselves against the cold, but paired off with canvas "sneakers." They seldom had gloves or mittens and depended upon tucking their hands into the sleeves of their coats to keep them warm. Signal communications were primitive in the extreme. Commonly the Chinese used the SCR-300, captured from the Chinese Nationalists, as their back-packed radio, the same radio used by the Marine infantry. Radio nets almost never went below the regimental level. Telephone wire was seldom strung beneath the battalion level. Below the battalion, communications was by runner supplemented with bugles, whistles, flares, and flashlights.

Lacking adequate communications at the front, Chinese attack patterns tended to be rigid and repetitive. Once committed, a Chinese battalion would usually stay in contact until completely shredded by casualties or until all its ammunition was used up. There was little or no battlefield resupply.

Lin Piao had been concerned over the capability of the poorly equipped Chinese to fight the Americans, but Peng Dehuai hammered home to his senior subordinate officers his belief that Americans were afraid of close combat, a tactic in which the Chinese Communist troops excelled. Peng himself was a specialist in what the Chinese called a "short attack," hammering away at enemy defenses with successive compact combat groups, usually not more than a company in size, until a breakthrough or puncture was achieved, a tactic not unlike that used by German storm troops in the last years of World War I.

U.S. Marines' and soldiers' imaginations sometimes magnified what they saw and heard while under attack. The Western press was soon filled with fantasies of "human sea attacks" by "hordes" of Chinese. Chinese propaganda photographs and films showing wave after wave of Chinese advancing in line across the snow with bravely flying red banners reinforced these exaggerations. The truth was quite different. Hearing or reading such reports, the Marine infantry, those who were really there, would later ask derisively: "How many hordes are there in a Chinese platoon?"

RCT-31 East of the Reservoir

In mid-afternoon on 26 November, Colonel MacLean and his command group arrived at Faith's position on Hill 1221. Faith, ignoring Murray's caution, received MacLean's permission to move his battalion forward the next morning to the position vacated by Taplett's battalion.

MacLean set up his regimental command post in a schoolhouse in Hudong-ni, a village about a mile south of Hill 1221. A big, robust, aggressive man, MacLean was 43, a graduate of West Point, Class of 1930, and a veteran of the European theater. Barr had given him command of the 31st Infantry about two months earlier, replacing a commander who had not done well in the Inchon to Seoul drive. Before that MacLean had been in the G-3 Section of the Eighth Army. Previously, in Japan, he had commanded the 32d Infantry and he knew Faith well.

5th Marines' 27 November Attack

Of the 1st Marine Division's planned attack to the west, Ray Murray later said: "It was unbelievable. The more you think about it, the more unreal it becomes. Well, anyhow, those were the orders and that's what we started to do."

All elements of Murray's RCT-5 were to be relieved by Monday noon, 27 November, so as to take positions at Yudam-ni preparatory to passing through RCT-7 to lead the advance to Mupyong-ni. First objective for the regiment, once it was altogether, was to be the road junction at Yongnim-dong, 27 road miles to the west.

By nightfall on 26 November, Roise's 2d Battalion, 5th Marines, was in its attack position at Yudam-ni. His company commanders gathered in his blackout tent at 2200 to receive the attack order. Two Corsairs from VMF-312 for close support and a "Grasshopper" from VMO-6 for aerial reconnaissance were promised. The 7th Marines would support Roise's attack with patrols and a secondary attack to the southwest. The temperature at Yudam-ni during the night went down to zero degrees Fahrenheit.

In the morning Fox Company, under Captain Uel D. Peters, led off the 5th Marines' attack with an advance up the road leading westward. Peters' first objective was a spur about 500 yards beyond the 7th Marines perimeter. Almost immediately his Marines were engaged by long-range small arms fire. The VMO-6 spotter plane, overhead as promised, reported Chinese positions all across the front. At 1115, Corsairs from VMF-312 dumped rockets and bombs on the Chinese emplacements in front of Fox Company. As Peters began his assault, Chinese soldiers could be seen fleeing to the west. Three prisoners were taken.

Dog Company, under Captain Samuel S. Smith, had followed behind Peters and at about noon joined in the fight. Altogether Roise's battalion advanced about a mile. At 1430, Roise ordered Peters and Smith to break off the attack and set up night defensive positions.

The 3d Battalion, 7th Marines, in its attack to the southwest, had advanced about the same distance, about a mile, before running into stiffening opposition. The battalion had a new commander: Lieutenant Colonel William F. Harris, 32, Naval Academy 1936, who had taken over from Major Roach on 11 November. He was the son of Major General Field Harris. As a captain he had been serving with the 4th Marines when it was surrendered to the Japanese on Corregidor in the Philippines. He had spent the war as a prisoner of war and was one of four former prisoners to witness the Japanese surrender on board the battleship *Missouri* (BB 63). A big man with an easy manner he was immediately liked by the Marines in his battalion.

At noon Taplett's 3d Battalion, 5th Marines, arrived at Yudam-ni, after a hard five-hour motor march from the east side of the reservoir, and was assigned an assembly area

Marines, probably members of the 5th Marines, take a roadside break while on the march from Hagaru-ri to Yudam-ni on 27 November. This photo shows very clearly the nature and condition of the MSR or "main supply route" that was the division's lifeline from Yudam-ni back to Hungnam.

National Archives Photo (USN) 80-G-424585

west of the village where the road forked to the north and west. Taplett understood that his battalion was to follow Roise's 2d Battalion when the attack was resumed in the morning.

The 1st Battalion, 5th Marines, now under Lieutenant Colonel John Stevens, did not arrive until dusk and was given an assembly area east of the village. Meanwhile, the 2d Battalion, 7th Marines, was completing its motor march, company by company, but without Randolph Lockwood, the battalion's new commander, who stayed behind in Hagaru-ri.

Units of the 5th and 7th Marines were now thoroughly intermixed and would become more so, but there was no specific jointure of command. Brigadier General Craig, the assistant division commander, might have been given command of the two regiments combined into a task force, but he was home on emergency leave. Colonel Litzenberg was much senior to Lieutenant Colonel Murray, and perhaps Smith thought that was all the overall command authority needed. Litzenberg had positioned his command post for the 7th Marines in the center of Yudam-ni. Murray's command post for the 5th Marines was some distance away in the northwest corner of the village.

During the day Almond, accompanied by an aide and an assistant operations officer, drove by jeep to Yudam-ni from his command post at Hamhung. Arriving at the 7th Marines command post unexpectedly, he found Litzenberg absent but his executive officer, Lieutenant Colonel Frederick R. Dowsett, 39, present. Tall, lanky, Dowsett briefed him on the enemy situation and the disposition of the regiment. Almond passed out three Silver Stars, one to an officer and two to enlisted Marines, and then late in the afternoon began his return to Hamhung. The MSR was jammed with traffic going in both directions. In his opinion, the traffic was poorly controlled. The drive took nearly five hours. That night he reported to GHQ in Tokyo that the strength of the enemy was considerable and that the disposition of the Marines needed to be reexamined.

Hagaru-ri, 27 November

At Hagaru-ri, Ridge's 3d Battalion, 1st Marines, completed the relief of Lockwood's 2d Battalion, 7th Marines, on the morning of 27 November. Companies D and E of Lockwood's battalion had already arrived at Yudam-ni. While they waited for their own battalion commander, Litzenberg attached the two companies to Davis' 1st Battalion.

Lockwood, in accordance with Smith's directive to Litzenberg, now led forward his remaining rifle company, Fox Company, to occupy Toktong Pass. He gave Captain William E. Barber orders to move off the road, beginning four miles north of Hagaru, with the mission of keeping open three miles of the MSR. Lockwood then returned to Hagaru-ri where his Headquarters Company and the remainder of his Weapons Company were awaiting trucks to take them on to Yudam-ni. The trucks never came. Lockwood himself, and the remainder of his battalion, would never get to Yudam-ni.

When Captain Barber took command of Fox Company on 7 November, he made a little speech, telling his company that he was "an infantryman and a hell of a good one at that." Born in Kentucky in 1919, he had enlisted in the Marine Corps in 1940. He went through parachute training, doing so well that he stayed on as an instructor. He was commissioned in 1943, and as a platoon leader at Iwo Jima with the 26th Marines, he was wounded and evacuated. Refusing to stay hospitalized, he came back to take command of a company. For this he received a Silver Star and his first Purple Heart.

Ridge, faced with the mission of defending Hagaru with two-thirds of a battalion, sent Major Joseph D. Trompeter, his S-3, and Major Edwin H. Simmons, his Weapons Company commander and supporting arms coordinator, on a walking reconnaissance. Trompeter and Simmons found that to enclose all of Hagaru-ri would require a perimeter of four miles, an impossible task for a single infantry battalion at two-thirds strength. Ridge estimated that one to two regiments would be required for a thorough defense.

"Under the circumstances and considering the mission assigned to the 1st Marine Division," General Smith would later comment, "an infantry component of one battalion was all that could be spared for the defense of Hagaru," adding with the benefit of hindsight, "This battalion was very adequately supported by air, and had sufficient artillery and tanks for its purposes."

Captain Benjamin S. Read's How Battery, 3d Battalion, 11th Marines, which had been shooting for the 2d Battalion, 7th Marines, was already in place in the northeast corner of the sketchy perimeter. Now it would have to divide its fire missions between the defense of Fox Company in Toktong Pass and the Hagaru-ri perimeter and at the same time provide its own defense for its segment of the perimeter. "Our lives centered on our 105mm howitzers, and our

Photo by Cpl L. B. Snyder, National Archives Photo (USMC) 127-N-A4644

Capt William E. Barber took command of Fox Company, 2d Battalion, 7th Marines, on 7 November. A seasoned combat leader, he had commanded a platoon and company at Iwo Jima. He demanded a lot from his Marines and he got it. Shortly before leaving Hagaru-ri to take position in Toktong Pass, he held a rifle inspection for his company.

mission was to support the infantry," said Captain Read crisply a short time later.

Captain Andrew J. Strohmenger's Dog Battery, 2d Battalion, 11th Marines, had arrived at Hagaru-ri with Ridge's 3d Battalion, 1st Marines. The battalion and battery had worked together before, notably at Majonni, and were old friends. Strohmenger's battery went into position on the flats just southeast of the village.

The extreme cold affected the recoil systems of the howitzers and the reach of their shells. The guns were slow in coming back into battery and the extreme range was cut down from 12,200 yards to something like 9,000-9,500 yards.

Not being able to be strong everywhere, Ridge decided to concentrate his two rifle companies, How and Item, in a salient southwest of the not yet operational, but all-important, airstrip. The other greatest threat to Hagaru-ri was the hill mass just east of the town that would come to be called "East Hill."

Beyond the airstrip, First Lieutenant Joseph R. "Bull" Fisher's Item Company improved the positions vacated by Barber's Fox Company by blasting deeper foxholes with "Composition C" plastic explosive. On Fisher's left flank, Captain Clarence E. Corley's How Company extended the line until it tied in with the right flank of Strohmenger's Dog Battery, 11th Marines. The frozen marsh in front of Dog Battery was covered with fire but left unmanned. The perimeter picked up again with a roadblock held by a portion of Weapons Company across the road running south to Koto-ri. East Hill remained unoccupied. Ridge planned to put George Company on the hill when it arrived from the south. Service Battalion held the roadblock on the road that led northeast of the reservoir. Somewhere out there on the east side of the reservoir was the Army column that would come to be called "Task Force Faith," named for its doomed commander. The rest of the perimeter was patched together with bits and pieces of the Service Battalion, the division's

A 105mm fires a mission from its position close to the airstrip at Hagaru-ri. The two batteries of artillery—Capt Benjamin S. Read's Battery H, 3d Battalion, and Capt Andrew J. Strohmenger's Battery D, 2d Battalion, both of the 11th Marines—were essential to the defense of the Hagaru-ri perimeter.

Photo by TSgt V. Jobs, National Archives Photo (USMC) 127-N-A130286

Headquarters Battalion, and odds and ends left behind by the 7th Marines, until it closed again on Item Company's right flank. At the northern-most edge of the perimeter, Read's How Battery, 11th Marines, like Strohmenger's battery, was used as a frontline unit.

Lockwood received orders from Litzenberg to move to Toktong Pass to assist Fox Company. He borrowed a platoon from Ridge's battalion as an escort, but the effort went nowhere. Tank-infantry patrols sent out to the north toward Yudam-ni and to the south toward Koto-ri were pushed back in by mid-afternoon.

East of the reservoir, Monday morning, 27 November, Colonel MacLean, commanding RCT-31, went forward, accompanied by Lieutenant Colonel Faith, and together they inspected the lines vacated by the Marines. MacLean then selected a forward command post site south of Faith's intended new position.

Chinese Order of Battle

It was not yet known with certainty, but the scattered Chinese elements encountered earlier by Murray's 5th Marines were from the *80th Division* of the *27th Army, Ninth Army Group*. Commanded by Sung Shih-lun, the *Ninth Army Group*, with a total of 11 and possibly 12 divisions, consisted of three "armies," the *20th, 26th,* and *27th,* each roughly equivalent to a U.S. corps in frontline infantry strength. Sung Shih-lun was the equivalent of a lieutenant general, but the Chinese Communist Forces had not yet adopted Western military grades. Rank was indicated by billet held. Sung, like Peng, was his own political commissar. The *Ninth Army Group* had been poised to invade Taiwan after having captured Shanghai from the Nationalists. At Mao's direction Peng had brought Sung up from the Shanghai area and had sent him into Korea with specific orders to destroy X Corps. Peng's headquarters, it will be remembered, estimated that Sung could bring 150,000 troops against 90,000 men, a close guess at the strength of X Corps, giving him a 1.7 to 1 advantage.

Mao, in a telegram sent to Peng on 12 November, said: "It is said that the American Marine First Division has the highest combat effectiveness in the American armed forces." Sung would make the destruction of the 1st Marine Division, as the strongest of the American divisions, his main effort.

Sung's information as to the location of Marine Corps units was excellent. His plan, as later pieced together by U.S. intelligence, was as follows: The *27th Army*—except for the *80th Division,* which was to come down the east side of the reservoir—was charged with attacking the two Marine regiments at Yudam-ni. The *20th Army* was to cut the MSR or main supply route south of Yudam-ni, including attacks against Hagaru-ri and Koto-ri. The *26th*, initially in reserve, would not come into the fight until somewhat later. Sung was to launch his attack the night of 25 November, simultaneous with the assault to the west against the Eighth Army, but he was not quite ready and he secured Peng's approval to delay his attack for two days.

Early on the afternoon of 27 November, Faith completed the move of his battalion into the positions vacated by Taplett's 3d Battalion, 5th Marines. It was a typical Marine Corps perimeter, horseshoe shaped and occupying the high ground. Each of the exposed sides was occupied by one of Faith's rifle companies, the battalion command post was in the center, and the open side to the rear was covered by elements of his Headquarters and Service Company and Weapons Company. Lacking the strength in men and weapons of a Marine battalion, Faith could not fill all the foxholes.

MacLean, who had returned to Hudong-ni, was told that several hundred Chinese had been sighted east of the Pungnyuri-gang inlet. He sent out his Intelligence and Reconnaissance Platoon to investigate. The platoon roared out of the compound in its machine gun mounted jeeps and was never seen again.

The 3d Battalion, 31st Infantry, commanded by Lieutenant Colonel William R. Reilly, arrived that

Chinese Order of Battle

Ninth CCF Army Group

20th CCF Army	*26th CCF Army*	*27th CCF Army*
58th CCF Division	*76th CCF Division*	*79th CCF Division*
59th CCF Division	*77th CCF Division*	*80th CCF Division*
60th CCF Division	*78th CCF Division*	*81st CCF Division*
89th CCF Division	*88th CCF Division*	*90th CCF Division*

Source: Montross & Canzona, *The Chosin Reservoir Campaign* (1957). The fourth division in each army was not organic. The *88th, 89th, and 90th CCF Divisions* were attached from the *30th CCF Army*. Some other, later, authorities, Chinese as well as American, show the *90th CCF Division* as the *94th CCF Division* from the *32d CCF Army*.

National Archives Photo (USMC) 127-N-A4886

LtCol Randolph S. D. Lockwood, Commanding Officer, 2d Battalion, 7th Marines, had allowed his headquarters to become separated from his rifle companies. On 27 November, on orders from Col Litzenberg, he attempted to reach Fox Company in Toktong Pass with elements of his H&S Company and Weapons Company shown here. The effort failed.

afternoon, followed by the 57th Field Artillery Battalion, commanded by Lieutenant Colonel Raymond Embree. MacLean put Embree's battalion—which was minus its Battery C—into a bivouac area near the hamlet of Sinhung-ni, just south of Pungnyuri-gang inlet. The two firing batteries were positioned on the south side of the inlet on low ground surrounded on three sides by ridges. Embree placed his artillery headquarters a mile or so farther south on the slope of Hill 1456. Battery D of the 15th Antiaircraft Automatic Weapons Battalion, with four full-track M19 weapons carriers mounting dual 40mm guns and four half-tracked M16 carriers bearing quad .50-caliber machine guns, was set up close to Embree's headquarters.

The 31st Heavy Mortar Company, with its 4.2-inch mortars, moved into a position close to MacLean's forward command post and about halfway between Faith's battalion and Reilly's battalion. Meanwhile, the 31st Tank Company, with 20 M-4A4 Sherman tanks and two 105mm howitzer tanks, had reached Hudong-ni.

Thus, on the evening of 27 November, elements of MacLean's RCT-31 were stretched out on the road for 10 miles in seven different positions. By nightfall, or shortly thereafter, Faith, on the northern end with his 1st Battalion, 32d Infantry, had registered his artillery and mortar defensive fires. At about this time he received orders from MacLean to attack the next morning toward Kalchon-ni. MacLean himself spent the night at Faith's headquarters.

Sung Shi-lun, it will be remembered, had allocated his *80th Division* to the attack east of the reservoir. Shortly before midnight a firefight developed on Company A's front on the forward edge of Faith's position. The company commander was killed. Stamford, the Marine captain, took temporary command. The Chinese attack spread until it encompassed the rest of the battalion perimeter.

South of the inlet, the two firing batteries of Embree's 57th Field Artillery Battalion and Reilly's 3d Battalion, 31st Infantry Regiment, came under heavy attack from the east. The Chinese overran the 3d Battalion's command post and both artillery batteries. Reilly was severely wounded. Farther south, mortar shells began falling on Embree's artillery headquarters. Embree, in turn, was badly wounded.

Yudam-ni, 27 November

As darkness fell on the 27th at Yudam-ni, Captain Wilcox's Company B, 7th Marines, which had been patrolling South Ridge, came under heavy attack. Lieutenant Colonel Davis, commanding the 1st Battalion, received permission from Litzenberg to take a company to extricate Wilcox. Davis led Charlie Company, less one of its rifle platoons and commanded by Captain

John F. Morris, down the MSR to positions across the road from Hill 1419. Baker Company pulled itself loose from its engagement and Davis took it back into Yudam-ni, leaving Morris' Charlie Company to occupy Hill 1419—about two miles south of the incomplete perimeter. With less than a full company, Morris organized a crescent-shaped defense on an eastern spur of Hill 1419, well below the crest.

Unknown to Litzenberg and Murray as yet was that almost surrounding them at Yudam-ni were the *79th* and *81st CCF Divisions.* Furthermore, the *59th CCF Division* had begun a wide enveloping movement past South Ridge and on south to cut the MSR at Toktong Pass, held only by Fox Company, 7th Marines.

Artillery support at Yudam-ni was provided initially by Major Parry's 3d Battalion, 11th Marines—three batteries of 105mm howitzers, 18 tubes in all, enough to support a regiment in a narrow zone of attack, but not enough to provide adequate 360-degree support for a sprawling two-regiment defensive sector. Fortunately, among the Marine forces converging on Yudam-ni, during that busy 27th of November, was the 4th Battalion, 11th Marines, commanded by Major William McReynolds, with three batteries of its heavier 155mm howitzers—18 more tubes. All would be in action before midnight.

Toward the end of the day on 27 November, two Marines at Yudam-ni help a wounded comrade reach an aid station. Heavy action at Yudam-ni had begun that morning when elements of both the 5th and 7th Marines made an attack westward, were halted, and fell back to defensive positions.

National Archives Photo (USMC) 127-N-A4904

That night the temperature dropped to 20 degrees below zero. Northwest Ridge, the last ridge to be occupied, now had a Marine presence, a frontline of foxholes chipped out of the frozen ground and occupied by tired and cold-benumbed Marines. How Company, 7th Marines, commanded by Captain Leroy M. Cooke, held Hill 1403, the high point on Northwest Ridge. On How Company's left flank were Easy and Fox Companies of Roise's 2d Battalion, 5th Marines, occupying the rest of the ridge until it dropped down to the defile through which passed the road to the west. Roise had his command post behind the juncture of these two companies. A roadblock manned largely by Weapons Company covered the road westward. On the other side of the road, Dog Company curled back toward Southwest Ridge.

Taplett, uneasy with the situation, turned the assembly area assigned his 3d Battalion, 5th Marines, into its own defensive perimeter. His command post was in a draw behind Hill 1282. He sent a platoon from Item Company to outpost a spur of Hill 1384 about 500 yards forward of his command post. The outpost began receiving harassing fire at 2045.

The *89th CCF Division's* attack against Northwest Ridge, with two regiments, the *266th* and the *267th,* began at about 2200. The Chinese suddenly hit all along the line with sub-machine guns and grenades supported by machine gun fire and an intense mortar bar-

Sketch by Sgt Ralph Schofield, USMCR

This BAR-man, as sketched by combat artist Sgt Schofield, sights in his Browning automatic rifle. It still has its bipod, making it an efficient substitute for a light machine gun. Some BAR-men threw away their bipods to lighten their load, reducing the BAR to an assault rifle.

rage. This attack apparently aimed at fixing the Marines in position while a dense column of Chinese assaulted the line on a narrow front against the boundary between Easy and Fox Companies. This assault penetrated the Marines' position, overrunning Fox Company's right flank platoon. Captain Samuel Jaskilka (a future four-star Assistant Commandant), commanding Easy Company, turned back his left flank to cover the penetration. Roise pounded the Chinese salient with his 81mm mortars and sent up a platoon from Dog Company to reinforce Fox Company's ruptured right flank. A great number of Chinese were killed and by dawn the break in the line had been repaired.

Things went less well in the fight that had begun for possession of Hill 1403. Captain Cooke, How Company's commander, had deployed his three rifle platoons in a semi-circle on the forward edge of the crest of the hill. His right flank crumbled under the weight of an assault by the *266th CCF Regiment*. Cooke himself bravely led a counterattack to restore his line and was cut down by Chinese machine gun fire. Second Lieutenant James M. Mitchell took temporary charge of the company until First Lieutenant Howard H. Harris, who had earlier commanded the company, could get there from battalion. Harris (no relation to Lieutenant Colonel William Harris, commanding the 3d Battalion) arrived at about midnight and found only one How Company officer, Second Lieutenant Minard P. Newton, Jr., still on his feet. The Chinese again assaulted Hill 1403 at about 0300. After an hour of pounding, Lieutenant Colonel Harris ordered the battered How Company to withdraw to the rear of the 2d Battalion, 5th Marines, leaving Hill 1403 in Chinese hands.

Battle for North Ridge

Concurrently with the assault of Northwest Ridge, the *79th CCF Division*, with three regiments, had moved against North Ridge, held by two widely separated companies of the 2d Battalion, 7th Marines—Dog Company on Hill 1240 and Easy Company on Hill 1282. Separating the two hilltops was a long saddle. (The battalion's third rifle company, Fox Company, it will be remembered, had been dropped off at Toktong Pass and Lockwood and his headquarters were still at Hagaru-ri.)

The *235th CCF Regiment* attacked in a column of battalions against Hill 1282 at about midnight. Easy Company, under Captain Walter Phillips, held its ground against the first attack. Simultaneously, the *236th CCF Regiment,* following behind the *235th,* was feeling out Dog Company's position on Hill 1240.

Anticipating an attack against North Ridge, Murray had moved Stevens' 1st Battalion out of its assembly area northward to the reverse slope of Hill 1282. First elements of Able Company reached a spur of Hill 1282 barely in time to reinforce Easy Company, 7th Marines, which was being pummeled by the *1st Battalion, 235th CCF Regiment.* Easy Company's commander, Captain Phillips was killed and would receive a posthumous Navy Cross. His executive officer, First Lieutenant Raymond O. Ball, took over command, was several times wounded, and died in the battalion aid station. Command devolved upon the senior platoon leader, First Lieutenant Robert E. Snyder. Easy Company had been reduced to the size of a single platoon, and by daylight the Chinese had taken the crest of Hill 1282.

The crest of Hill 1240 to the east had also fallen. Chinese from the *3d Battalion, 236th CCF Regiment,* had overrun the command post of big, burly Captain Milton Hull, the company commander of Dog Company. At about 0300, Hull, wounded, counterattacked with the few squads at his disposal, won back a foothold, and was wounded again. When dawn came he could count only 16 Marines left with him, and the enemy had him surrounded.

During the night some Chinese had crossed the saddle that separated the Dog and Easy Company positions and had taken the 5th and 7th Marines' command posts under fire. Some time before midnight, a few half-dressed mortar men from How Company, 7th Marines, beaten back from Hill 1403, found their way into Taplett's 3d Battalion, 5th Marines, perimeter. A message from How Company, 7th Marines—that part that still remained on Hill 1403—reached Taplett, warning him that the Chinese were flanking his position. At about 0145, Taplett's outposted platoon on Hill 1384 received increasingly heavy fire. Shortly thereafter a CCF force, estimated at two companies, overran the outpost. Taplett's command post became the bull's eye of the fight. Major John J. Canney, the battalion executive officer and a World War II aviator turned infantryman, was killed.

South of Yudam-ni

At 0230, with the assaults against North and Northwest Ridges at their height, the Chinese also struck Charlie Company, under Captain Morris, on the spur of Hill 1419 two miles south of Yudam-ni. Morris' Marines held on grimly until dawn when artillery fire finally made the Chinese break off their attack. But, with a third of his men casualties, Morris was effectively pinned into position by Chinese fire continuing to rain down from the heights. His Marines could do nothing more than hold their position and hope that help would come from Yudam-ni.

While the *79th* and *89th CCF Divisions* savaged the Marines on Northwest and North Ridges, the *59th CCF Division* completed its wide sweeping movement to the southeast, putting itself in position to cut the 14 miles of vital MSR between Yudam-ni and Hagaru-ri. Until midnight on 27 November truck traffic on the MSR was still active and unimpeded—mostly empty trucks from Lieutenant Colonel Beall's 1st Motor Transport Battalion rattling their way back to Hagaru-ri, having delivered the last serials of the 1st Battalion, 5th Marines, and the 4th Battalion, 11th Marines, to Yudam-ni.

Captain Barber had gone into position at Toktong Pass with a near full-strength Fox Company reinforced with sections of water-cooled Browning machine guns and 81mm mortars from Weapons Company, 2d Battalion—a total of 240 officers and men. Barber chose to organize his defensive perimeter on a hill at the mid-point of the pass. "We arrived in the late afternoon after which we unloaded and were positioned for the night," remembered Corporal Howard W. Koone. "Our position was off to the right of the road up on a saddle-like hill. The ground was like a sheet of concrete and very barren."

Barber's 3d Platoon, under First Lieutenant Robert C. McCarthy, occupied the high ground at the center of the narrow perimeter. At about 0230, McCarthy's two forward squads were overwhelmed by a company-sized attack. Out of 35 men, McCarthy lost 15 killed, 9 wounded, and 3 missing. The eight survivors fell back to the reserve squad on the reverse slope of the hill. Barber's position was almost cut in half, but his two wing platoons managed to hold their ground. Much was owed to the valor of three Marines: Private First Class Robert Benson and Private Hector A. Cafferatta of the 2d Platoon under Second Lieutenant Elmo G. Peterson on the left, and Private First Class Gerald J. Smith of the 1st Platoon under First Lieutenant John M. Dunne on the right. One party of Chinese penetrated as far as the company command post and the 81mm mortar position. Fighting, some of it hand-to-hand, continued until daybreak when the Chinese broke off the assault but continued to keep the position under fire. In all Barber had lost 20 Marines killed and 54 wounded. Fox Company did not know how many Chinese it had killed but guesses went up to 500.

Howard Koone was one of those wounded. He eventually found himself in a Korean hut being used by Fox Company's corpsmen as a sick bay. He was told that helicopters would be coming to evacuate the wounded and that he would be third on the list, but the helicopters never came.

Yudam-ni, 28 November

Dawn on 28 November saw the tactical situation on Northwest Ridge unresolved. Hill 1403 had been lost to the enemy. Elsewhere both Marines and Chinese were clinging to the high ground. Roise's 2d Battalion, 5th Marines, had a firm grip on its portion of the line. As yet there had been no orders to abandon the offensive begun the day before. Roise had received orders from Murray to continue the attack at daybreak. Taplett's 3d Battalion was to come up on his right flank and add its weight to the assault.

Murray met with Litzenberg at the 7th Marines command post at dawn. Both regimental commanders agreed that the situation dictated that they change from the offensive to the defensive. Murray canceled the attacks to be made by Roise and Taplett.

Murray barely knew Litzenberg. In the south, at Seoul, he had seen him once or twice at division headquarters. The only intimate contact he ever had with him would be at Yudam-ni. Never-the-less, the loose command relationship seemed to work. "If he had troops on some hills," said Murray, " then I put troops on some other hills, so that we had a good perimeter defense of the area."

Murray remembered that Litzenberg "had a reputation of being sort of a fussbudget, a stickler . . . he seemed to be a studious type of person, knew his business, and as far as I could tell from talking with people in the 7th Marines, it seemed everyone respected him and his abilities. . . . Many people have asked why he didn't just assume command up there. I can't answer that question definitively. After all, there was a division headquarters over the hill from us, and we were still part of that division, so we had a common head. But in any case, we decided to operate very closely together, and we did."

Taplett had begun his counterattack against the spur of Hill 1384 at about 0300 with two platoons of George Company led by Lieutenants John J. "Blackie" Cahill and Dana B. Cashion. Some time after daylight Cahill and Cashion reached the crest of Hill 1384 with their platoons. About this time Taplett received the order canceling the attack. He, in turn, directed Cahill and Cashion to hold where they were until they received further orders. With their presence on top of the hill, the remainder of How Company, 7th Marines—some 80 officers and men—was able to complete its withdrawal from Hill 1403 and pass on into Taplett's perimeter.

John Stevens' 1st Battalion, 5th Marines, spent the morning consolidating its position on Hill 1240. His Charlie Company, under Captain Jack R. Jones, had moved over during the night to backstop Taplett's battalion and was put under the operational control of the 7th Marines. One platoon was dropped off to rejoin its parent

A row of dead Chinese, frozen in grotesque positions, on the high ground overlooking a command post of the 5th Marines at Yudam-ni, mark the line of their farthest advance. The burden of the Chinese attack was borne chiefly by isolated Marine rifle companies holding ridgeline positions.

Photo by Sgt Frank C. Kerr, Department of Defense Photo (USMC) A4839

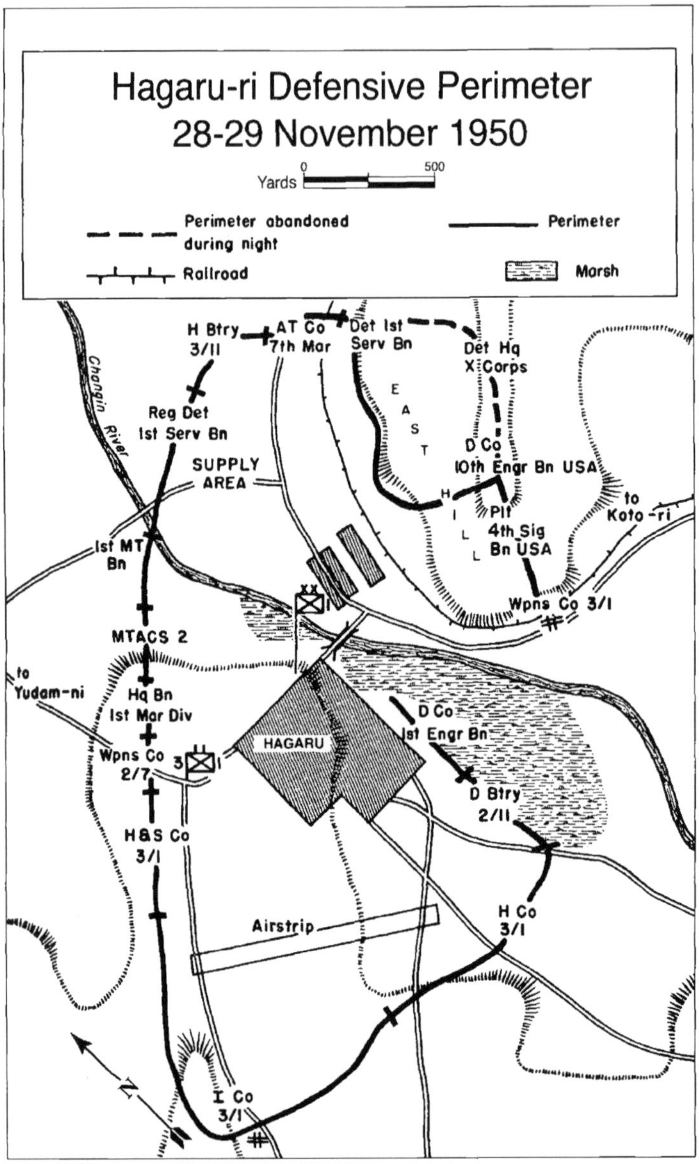

battalion on Hill 1240. The remainder of the company continued on to Hill 1282. A company of Chinese from the *235th CCF Regiment* had lodged itself on the hilltop. Jones led his reduced company in a hand-to-hand assault that won back the hill.

At 1100 Murray ordered Roise to pull his battalion back to Southwest Ridge tying in with Harris' 3d Battalion, 7th Marines, on his left and Taplett's 3d Battalion, 5th Marines, on his right. Early that afternoon Roise brought his battalion back from Northwest Ridge a company at a time. Except for occasional harassing fire, the Chinese did not interfere.

Early that morning, Davis' 1st Battalion, 7th Marines, had set out to the south to relieve both Charlie and Fox Companies from their encirclement on the MSR. Able Company, under First Lieutenant Eugenous M. Hovatter, led off, moving through a gorge separating South from Southeast Ridge. Five hours of fighting found Able Company still a mile short of Charlie Company's position. Baker Company, under Captain Wilcox, joined the attack. Together the two companies reached Charlie Company. Litzenberg, with Charlie Company now relieved and its wounded evacuated, and not wanting to have the 1st Battalion trapped in the gorge, ordered Davis to pull back into the Yudam-ni perimeter. By evening Stevens' 1st Battalion, 5th Marines, had relieved the shattered remnants of Hull's Company D, 7th Marines. The two regiments at Yudam-ni, their perimeter tightened and mended, faced the night of 28 November with considerable confidence. But Barber's Fox Company remained alone in Toktong Pass.

Almond Visits Faith

At Hagaru-ri, a platoon-sized patrol, sent out to the southwest early on the morning of 28 November from Fisher's Item Company, was pushed back into the perimeter. At about the same time as this patrol action, Ridge telephoned Colonel Bowser, the division G-3 and Smith's war chief, recommending that an overall defense commander be designated for Hagaru-ri. He also requested that the arrival of his George Company and the Royal Marine 41 Commando be expedited. Before a decision could be reached, General Smith flew in by helicopter at about 1100 to open his command post at Hagaru-ri. A half-hour later General Almond, along with his junior aide, 26-year-old Captain Alexander M. Haig, Jr., arrived in Almond's L-17 light aircraft, the "Blue Goose."

After meeting with Smith, Almond borrowed a Marine helicopter to take him east of the reservoir to meet with Faith and MacLean. Colonel MacLean, it will be remembered, had spent the night of 27 November at Faith's position. MacLean thought that Faith's battalion had come through the night in fairly good shape. He knew little or nothing about what had happened south of the inlet. At dawn he left to return to his own advance command post. His short jeep trip was not interrupted.

Almond, on arriving at Faith's position in his borrowed Marine Corps helicopter, airily told Faith that there was nothing in front of him except scattered Chinese retreating to the north and that he should try to retake the lost high ground. As further encouragement, Almond informed Faith that he had three Silver Stars to present, one for Faith himself and two more for whomever Faith designated. Faith called forward a wounded platoon leader and a mess sergeant. Almond pinned the three Silver Stars to their parkas, Captain Haig noted their names in his notebook, and the general and his aide got back on board their helicopter. As the helicopter whirled away, Faith and the lieutenant tore the Silver Stars from their parkas and threw them in the snow.

Stopping to see MacLean, Almond advised him that the previously planned attack would be resumed once the 2d Battalion, 31st Infantry, joined the regiment. This battalion and Battery C of the 57th Field Artillery were marooned far south on the clogged MSR.

Photo courtesy of Sgt Norman L. Strickbine, USA

East of the reservoir during 28 November, the Army's 3d Battalion, 31st Infantry, and two firing batteries of the 57th Field Artillery Battalion painfully reorganized after the previous night's devastating attack against their positions just south of Pungnyuri Inlet. Bodies can be seen in the foreground. Sporadic fighting continued north of the inlet.

During the night, the 31st Medical Company, pushing north from Hudong-ni had been ambushed and badly shot up in the vicinity of Hill 1221. Survivors drifting back to the headquarters of RCT-31 at Hudong-ni were the first indication that the road had been cut.

Meanwhile, General Hodes, the assistant commander of the 7th Division, was at Hudong-ni. He directed Captain Richard E. Drake, commander of the 31st Tank Company, to sally forth to the north to see if he could break through to the inlet. Drake moved out with 16 tanks. Hodes rode with Drake as a passenger; he did not take tactical command. Without infantry support, the tanks could not break the Chinese grip on Hill 1221 which effectively blocked the route north. Four tanks were lost. Hodes returned to Hudong-ni in a jeep, intent on getting back to Hagaru-ri for help. He took a tank, at Drake's insistence, for transportation and got back to Hagaru-ri, five miles away, without further incident. He never returned to Hudong-ni.

South of the inlet that day, 28 November, the badly battered 3d Battalion, 31st Infantry, and 57th Field Artillery Battalion painfully reorganized and consolidated their positions. Before nightfall, the Chinese came back into the attack with the M16 and M19 self-propelled guns the focal point of their effort. The automatic 40mm and .50-caliber fire did its lethal work. The perimeter held and many Chinese died.

North of the inlet sporadic fighting had continued. A dominant hill position was lost to the Chinese. Stamford ran close air support strikes with Marine Corsairs with little apparent effect. To the east the battalion could glimpse long columns of Chinese marching south, some of them mounted on

Mongolian ponies, or so it was said. Air strikes were flown against them and claimed good results.

Hagaru-ri, 28 November

Smith had moved into a Japanese-style house, soon overcrowded with the impedimenta of a division command post. On the wall close by Smith's field desk hung a picture of Stalin; Smith let it remain where it was. By nightfall on the 28th Smith had officially sanctioned actions already taken at Yudam-ni. Murray was ordered to halt his attack to the northwest. Litzenberg was told to attack to the south and reopen the MSR to Hagaru-ri. Together, Murray and Litzenberg were to plan for the continued defense of Yudam-ni and the breakout to the south. The joint defense plan worked up by Litzenberg and Murray provided for RCT-5 to take over responsibility for the west and north sectors, RCT-7 for the east, south, and southwest.

"Although the two regimental commanders acted jointly," said Taplett years later, "I harbored the gut feeling that Colonel Litzenberg and not Colonel Murray called the shots simply because of seniority. I confess to having more confidence in Murray."

During the afternoon, Colonel Bowser, the division G-3, telephoned Lieutenant Colonel Ridge confirming his appointment as Hagaru's defense commander. By then Ridge knew that George Company would not be arriving in time to occupy East Hill. George Company under Captain Carl L. Sitter reached Koto-ri that same day. Sitter, 28, had received a field commission in World War II after two years enlisted service. He fought in the Marshalls and at Guam, was twice wounded, and had received a Silver Star. At Koto-ri it soon became obvious that Sitter's company could go no farther without strenuous effort.

Before moving north to the reservoir, the commander and staff of the division's artillery regiment, the 11th Marines, lined up outside their command post at Hamhung for a group photograph. Standing, from left to right, are: Maj Donald V. Anderson, LtCol Carl A. Youngdale, the executive officer, Col James H. Brower, the commanding officer, and LtCol James G. Appleyard. Kneeling are Capt William T. Phillips and Maj Floyd M. McCorkle.

Photo by TSgt James W. Helms, Jr., National Archives Photo (USMC) 127-N-A5264

Colonel Brower had arrived at Hagaru-ri with the headquarters of his artillery regiment, the 11th Marines. He set up the fire support control center in juxtaposition with Smith's headquarters. His executive officer, Lieutenant Colonel Carl A. Youngdale, 48, Iowa State University, class of 1936, headed the regiment's fire direction center.

That afternoon, Company D, 10th Combat Engineer Battalion, came in from a tent camp the engineer soldiers had set up just outside the perimeter on the road leading south to Koto-ri. In his expanded role as Hagaru-ri defense commander, Ridge had operational control of the company. He decided to use it to fill the yawning gap on East Hill. He so informed the engineer company commander, Army Captain Philip A. Kulbes. The engineer captain protested, saying that he was at Hagaru to build a new command post for X Corps and that his men—77 Americans and 90 South Koreans—had no training in infantry combat. Aside from individual weapons, the only armament the company possessed was four .50-caliber machine guns, five light .30-caliber machine guns, and six 3.5-inch rocket launchers. Ridge asked Kulbes if he would accept the tactical advice of a Marine officer and Kulbes said he would. Captain John C. Shelnutt, the executive officer of the 3d Battalion's Weapons Company, was assigned as a "liaison" officer. Shelnutt was accompanied by a radioman, Private First Class Bruno Podolak. Major Simmons privately advised Shelnutt that, in face of the Army captain's reluctance, he would have to take de facto command. The Army company procrastinated, taking its time to move its trucks and engineer equipment into a motor park. At dusk the engineers started up the hill. The ascent took them through the roadblock facing south toward Koto-ri.

About 10 Marines under Gunnery Sergeant Bert E. Elliott, the Weapons Company machine gun platoon sergeant, manned the roadblock. Reinforcements for the roadblock came late in the day in the form of a platoon from the Army's 4th Signal Battalion sent to install communications for what was to be General Almond's command post. The Army signal lieutenant, First Lieutenant John A. Colborn, like the Army engineer captain, reported that his men had no infantry training. The 3d Battalion's Weapons Company commander asked him if he would take orders from a Marine gunnery sergeant. The lieutenant eagerly said that he would.

On the north side of East Hill, the commanding officer of the 1st Service Battalion, Lieutenant Colonel Charles L. "Gus" Banks, 36, a World War II Edson's raider, was named a sub-sector commander. He was to coordinate his actions with Lockwood. While Kolbes' engineers climbed the south face of the hill a column of Marines sent by Banks started up the north face. The two columns were supposed to meet on the crest.

At sundown, Simmons pulled his roadblock back about 75 yards to what he thought was a stronger position. With a total of about 40 men Gunnery Sergeant Elliott was able to man the roadblock with his Marines and the knoll on his left flank, which was the first step in the climb up East Hill, with the Army signal platoon. Elliott had been a lieutenant during World War II and he was determined to win back his bars. Tough, battle-wise, and not particularly well-liked, even by his own Marines, he balanced his .45-caliber pistol in the palm of his hand and bluntly advised the soldiers that if they dug in, stayed, and fought, they would be there in the morning; but if they got up to run, he would shoot them himself.

Hagaru-ri Airstrip Defense

While the two columns, Army and Marine, moved toward the crest of East Hill, a major Chinese attack hit the southwest quarter of the perimeter, fortunately striking the strongest segment of the Marine line, that held by Companies H and I of the 3d Battalion, 1st Marines. The two companies, stretched thinly along the far side of the prospective airstrip on which the engineers were laboring under lights, were well dug-in. Holes had been blasted through the top 8 or 10 inches of frozen earth with ration cans filled with C-3 explosive so that foxholes and machine gun emplacements could be dug. The spoil was used to fill sandbags. A meager supply of concertina and other barbed wire was strung out where it would do the most good. Five-gallon cans filled with gasoline were rigged with white phosphorus grenades. Tied to the grenades were strings that could be pulled to explode the grenades and flame the gasoline. An earnest demolitions sergeant explained that these were French devices known as a "foo-gah-say." Three draws led into the Marine position; they had been sown with anti-personnel mines. In all, it would be a tough nut for the Chinese to crack.

A light snow was falling. The two companies were at 100 percent alert. At about 2230, three red flares and three blasts of a whistle signaled that the Chinese were coming. Mortar shells, high explosive mixed with white phosphorus, began crunching down on the frontline positions. Marine supporting arms—some artillery but mostly mortars and machine guns—took the Chinese under fire, but did not stop the enemy from closing to within hand-grenade and burp-gun range. The assault continued for an hour, the Chinese attacking in combat groups of about 50 men each. Most of the Marine line held, but the Chinese succeeded in penetrating the center of How Company's position. The company commander, Captain Clarence Corley, pulled together a scratch squad and tried to plug the gap but was pushed aside. Some few Chinese broke through as far as the airstrip where the engineers killed them.

Ridge dispatched a mixed platoon of Marines and soldiers under First Lieutenant Grady P. Mitchell, Jr., to back up How Company. Mitchell was killed and First Lieutenant Horace L. Johnson, Jr., took over. Johnson deployed his men in a ditch fortuitously behind How Company's ruptured line. The Chinese who had penetrated the position were milling around, seemingly more intent on looting the supply and cook tents than exploiting their success. They were fighting for food, warm clothes, and U.S. ammunition. At least one wounded Marine survived by feigning death when a Chinese soldier stripped him of his parka. Ridge fed in another platoon made up of casuals to build on Johnson's line. By about 0130 the situation appeared to be under control. The engineers relit their floodlights, got back on their dozers, and resumed work on the airstrip.

But bad things were now happening on the other side of the perimeter.

Action on East Hill

The two columns that had been sent up East Hill had failed to reach the crest. Captain Shelnutt, in virtual command of the Army engineers and under heavy fire, reported to the 3d Battalion's Weapons Company commander that Banks' column coming up the other way did not seem to be where it was supposed to be. Shelnutt was told to turn back his left flank and hold for the night. He was promised that artillery fire would fill in the gap.

At about 0115, the Marines and soldiers on the south roadblock were treated to the sight of a company-sized column of Chinese marching up the road toward them. Apparently the pullback of the roadblock earlier in the evening had caused the Chinese to think the position had been abandoned. The column presented the pair of Weapons Company water-cooled machine guns with a perfect enfilade target. Few members of the Chinese column escaped.

At Ridge's command tent heavy small arms fire and grenades could be heard on East Hill itself. The Weapons Company commander reached Shelnutt by radio at about 0200. Podolak, the radio operator, informed him that Shelnutt was dead: "There's nobody up here except me and a couple of doggies." Podolak was sternly enjoined, as a Marine, to take charge. The next time the Weapons Company commander tried to radio him the set was dead.

During the night, stragglers from the Army engineer company, mostly South Koreans, but some Americans, streamed back off the hill and took cover in the ditches and culverts of Hagaru-ri itself. Some few were rallied into a support line, stiffened with a handful of Marines, along the road paralleling the base of the hill. Other soldiers stayed on the hill and fought bravely. Most of these died.

Across the perimeter, Captain Clarence Corley, a spent bullet in his arm, launched a counterattack at about 0430 to restore his main line of resistance. It was successful, but the night had cost How Company 16 men dead and 39 wounded.

Hagaru-ri, 29 November

Ridge's greatest concern now was the situation on East Hill. If the enemy continued to have possession of the crest when daylight came, exposing the defenses of Hagaru-ri to full view, the situation would be critical. At 0530 he decided that he must counterattack. Major Reginald R. Myers, 31, University of Idaho 1941, the battalion executive officer, volunteered to lead a column up the hill. Myers had spent most of World War II on sea duty but joined the 5th Marines in time for Okinawa and North China. There was no tactical unit available to him at Hagaru that could be used. The attack would have to be made by a mixed force of service troops—and some stragglers found skulking in the town—patched together into a provisional company of about 250 men, mostly Marines but including a few soldiers. Myers' improvised company formed up on the road next to the battalion command post and was tolled off into platoons and squads. The first platoon, made up of Marines from the 1st Engineer Battalion and under command of First Lieutenant Robert E. Jochums, was the most homogenous and in the best shape.

Ridge delayed Myers' jump-off until about 0930 by which time the morning mists had cleared and Corsairs for close support could be brought overhead. The south roadblock had held. The soldier signalmen had stayed and fought well, delighting both themselves and the Marines. Myers led his "company" upward through their position. Troubles began almost immediately, if not from Chinese gunfire then from the icy slope. Men stumbled and fell, to be hauled to the rear by others only too willing to carry them to relative safety. Myers' force melted away to about 75 men. Best performance, predictably, was by the platoon led by Jochums. He was wounded in the foot but continued in command. Myers could claim reaching the military crest, but the topographical crest was still firmly in Chinese hands.

A supporting attack was to be made by Company A, 1st Engineer Battalion, under Captain George W. King, coming up the south face and passing through Myers' position. King started up the hill at about noon, his 1st Platoon under First Lieutenant Nicholas A. Canzona in the lead. Orders were changed. King's company was pulled back, marched almost a mile to the north, and then sent up the north face. Like the Myers force, he reached the military crest, but on the north side. His company went into a reverse slope defensive position for the night, separated from Myers by about 500 yards. Ridge had to be satisfied with King and Myers holding these positions. The Chinese continued to hold the topographical crest.

Ridge planned to feed in Sitter's George Company to take the remainder of the hill when it arrived from Koto-ri. From prisoner interrogation Ridge now believed that the *58th CCF Division*, led by the *172d Regiment* and followed by the *173d*, with the *174th* held back in reserve, had attacked Companies H and I. It was not clear what Chinese force was driving west to East Hill or when its deployment to assault Hagaru itself would be completed.

The Corsairs from Frank Cole's VMF-312 flew 31 sorties that day over Hagaru most of them against East Hill. One plane took a bad hit from Chinese small arms. The pilot, First Lieutenant Harry W. Colmery, successfully crash-landed inside the perimeter.

Brigadier General Hodes, the assistant division commander of the 7th Division, had spoken briefly with General Smith upon his arrival from Hudong-ni on the evening of 28 November. At noon on the 29th of November, he met again with Smith, informing him in more detail of the condition of RCT-31 east of the reservoir; that it had taken 400 casualties and was falling back toward Hagaru-ri and probably was unable to fight its way to safety.

LtCol Allan Sutter arrived at Koto-ri with the 2d Battalion, 1st Marines, on 24 November. Col Puller moved his headquarters to Koto-ri on the same day. With the 3d Battalion at Hagaru-ri and the 1st Battalion at Chinhung-ni, Puller would have his infantry battalions at three widely separated combat bases along the MSR.

National Archives Photo (USMC) 127-N-A5546

"The inference was that they should be rescued by a larger force," wrote Smith in his log. "I have nothing now with which to lend a hand except the battalion at Hagaru-ri and it has its hands full. I cannot see why the cutoff battalions cannot at least improve the situation by moving toward us."

Second Night on Fox Hill

Barber was supposed to have brought Fox Company off Toktong Pass and, with the help of Davis' battalion, was to have marched on into Yudam-ni. There was no chance of this. He was already encumbered with 54 wounded. During the morning of 28 November he had the help of a close air strike by Australian F-51 Mustangs. Later he sent out patrols that confirmed that he was completely surrounded. He asked for resupply by air. Marine R-5D four-engine transports, the Marine Corps equivalent of the Air Force C-54, dropped medical supplies and ammunition. Most fell at the base of the hill. Recovering them cost two more Marines wounded.

That night the Chinese came again against Fox Company. Five more Marines were killed, 29 more wounded, among the latter Captain Barber. Hit in the leg, he received first aid and stayed in action. During the day that followed, both Marine and Air Force planes dropped ammunition and other supplies. A Marine helicopter made a precarious delivery of some ammunition and much-needed radio batteries. Lieutenant Peterson, already twice wounded, took a patrol out in front of Fox Company to recover some errant mortar ammunition.

Koto-ri Action, 24-28 November

Lieutenant Colonel Alan Sutter's 2d Battalion, 1st Marines, had arrived at the meager village of Koto-ri on 24 November. Handsome, silver-haired Sutter, 36, Dartmouth 1937, had been a signal officer at Guadalcanal, Guam, and Okinawa. Now at Koto-ri, with his battalion reinforced by the 105mm howitzers of Easy Battery of the 11th Marines, a platoon of 4.2-inch mortars, bits and pieces of the regimental antitank company, and Company D of the 1st Medical Battalion, he set up a conventional perimeter defense.

A patrol from Captain Jack A. Smith's Easy Company brushed up against about 25 Chinese west of the village and brought in two wounded prisoners who said they were part of a Chinese division moving into attack positions. Chesty Puller and his regimental headquarters had joined Sutter at Koto-ri and in the next several days more Marine and Army units jammed their way into the protective envelope of the perimeter. On the morning of 28 November, Smith ordered Puller to send a force up the MSR to meet a tank patrol coming down from Hagaru-ri. Sutter sent out Dog Company under Captain Welby W. Cronk, but it was stopped a mile north of the perimeter by a strong Chinese force entrenched on both sides of the road. Dog Company withdrew under cover of air strikes by the busy Corsairs of VMF-312. The day's fighting cost the Marines four killed and 34 wounded. Three prisoners were taken and they identified their unit as the *179th Regiment, 60th CCF Division.*

Solemn Meeting at GHQ

The 28th of November had been a busy day for General Almond and, when he arrived at his comfortable headquarters at Hamhung that evening, he found urgent orders directing him to report immediately to MacArthur's GHQ in Tokyo. Almond and a small staff left for Tokyo from Yonpo in an Air Force C-54. They arrived at Haneda Airport at 2130 where Almond was told to proceed immediately to General MacArthur's residence at the American Embassy. He learned that MacArthur had called his senior commanders back to GHQ for a secret council of war. General Walker would also be present. The conference lasted two hours. In the west, Eighth Army's "Home-by-Christmas" offensive had gone well for the first two days. Then, on the night of 25 November, Chinese bugles were heard all across the front. By noon on 27 November, Walker had reported to MacArthur that he estimated there were 200,000 Chinese in front of him, that the ROK II Corps had been swept away, and that the U.S. IX Corps was falling back to cover his exposed flank. Walker now informed MacArthur that he thought he could build up a line in the vicinity of Pyongyang. Almond, in a bit of braggadocio, told MacArthur that he expected the 1st Marine and 7th Infantry Division to continue their attack. However, in his own mind Almond had come to realize that the greatest problem facing X Corps was its dispersion over a 400-mile front. He had begun to contemplate concentrating his

Photo by Cpl W. T. Wolfe, Department of Defense Photo (USMC) A4868

A company of Marines, quite possibly Capt Welby W. Cronk's Company D, 2d Battalion, 1st Marines, makes a school-solution advance up a hill outside the Koto-ri perimeter on 28 November. One platoon is deployed as skirmishers followed by the other two platoons in squad column. Cronk took three Chinese prisoners, but the Marines lost four killed and 34 wounded.

forces into a perimeter defense around the Hamhung-Hungnam area. MacArthur, after listening to his field commanders, gave his decision: a changeover from the strategic offensive to the defensive. (Some authorities believe MacArthur had already reached this decision before meeting with his senior field commanders.)

Yudam-ni, 29 November

The night of 28-29 November was quiet at Yudam-ni. Division directed that an effort again be made to relieve Fox Company. A composite battalion was pasted together of Able Company from the 5th Marines, Baker and George Companies from the 7th Marines, reinforced with a section of 75mm recoilless rifles and two sections of 81mm mortars. Major Warren Morris, the executive officer of 3d Battalion, 7th Marines, was placed in command. He assembled his force in front of the 1st Battalion, 7th Marines, command post on the morning of 29 November and at 0800 marched out to the south. Three hundred yards outside the perimeter heavy machine gun fire laced into his column. Morris pushed on. Corsairs came overhead to help and dropped messages warning that the Chinese were entrenched along both sides of the road. At 1315, Litzenberg, warned that Morris' column was in danger of being surrounded, ordered him to return to the perimeter. Fox Company would spend another night at Toktong Pass alone. As evening fell, Captain Barber called his platoon leaders together and told them they could expect no immediate relief.

RCT-31 Begins Withdrawal

During the night of 28 November, Faith and MacLean tried to get some sleep in the hut that was now their joint command post. By 0100, 29 November, the Chinese were attacking in strength but were beaten off. An hour later, MacLean ordered Faith to prepare to breakout to the south with the objective of reaching the 3d

Battalion, 31st Infantry. All trucks were to be unloaded of cargo and given over to carrying out the wounded. Equipment and vehicles left behind were to be disabled but not destroyed. Blackout would be observed and there would be no burning of tentage and supplies.

MacLean and Faith began their march-out at about 0600 on Wednesday morning, the 29th. It was strangely quiet as the rifle companies broke contact and came down from the high ground. The truck column, about 60 vehicles in all, formed up and moved south on the road with Marine Corsairs overhead. Leading the way was a command party that included MacLean and Faith. As the party approached the highway bridge over the inlet, it came under fire and split into two parts, MacLean with one, Faith with the other.

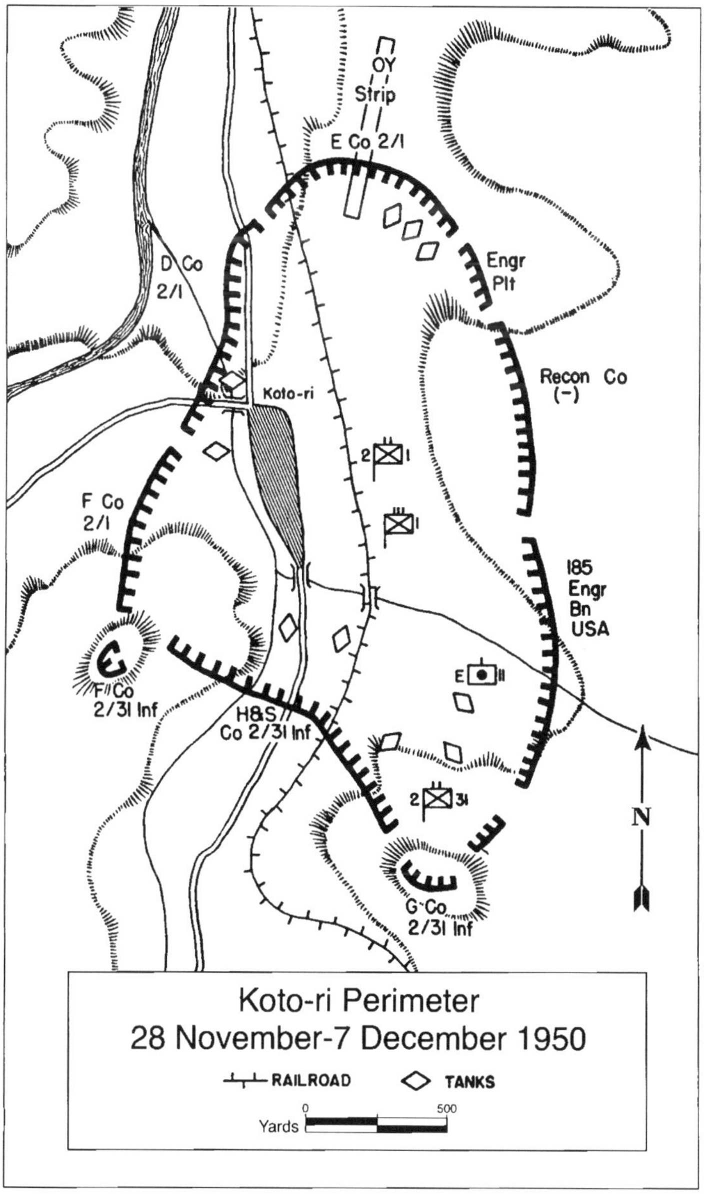

A column of troops was seen coming up the road. "Those are my boys," shouted MacLean and he started on foot across the ice toward them. A crackle of rifle fire was heard. His body was seen to jerk as though hit several times by bullets. He fell on the ice, then got to his feet and staggered on until out of friendly sight. Much later it would be learned that he was taken prisoner, but on the march north died of his wounds. His comrades buried him by the side of the road.

Faith was now the senior surviving officer and the 31st RCT would go into the collective memory of the Korean War as "Task Force Faith," although it would never officially bear that name. The head of Faith's column reached the 3d Battalion's positions by 0900 and by 1300 most elements had closed south of the inlet. Faith formed a new perimeter with the remnants of the two battalions, attempting to incorporate some of the high ground to the south. A helicopter sent in from Hagaru-ri by General Hodes took out the two wounded battalion commanders, Reilly and Embree. Air delivery of ammunition and supplies, called in by Stamford, had mixed results. Much of what was dropped landed outside the new perimeter.

Faith knew nothing of Drake's attempt to reach the inlet with his tanks. Drake tried a second time on 29 November with 11 tanks and a scratch platoon of infantry drawn from the regimental headquarters. After four hours of effort, the tanks fell back once more to Hudong-ni.

Fox's Continued Ordeal

The night of 29-30 November was again relatively quiet at Yudam-ni, but not so at Toktong Pass. At 0200 a voice came out of the dark and in stilted English said: "Fox Company, you are surrounded. I am a lieutenant from the 11th Marines. The Chinese will give you warm clothes and good treatment. Surrender now."

Fox Company threw up some 81mm illumination shells and replied with mortar and machine gun fire. The Chinese were caught in their attack position, perhaps three companies of them. Many died but some got close enough for an exchange of hand grenades. Fox Company, now well dug in, lost only one Marine wounded. At sunrise the protective Corsairs came overhead once again.

Chinhung-ni Action, 26-30 November

Short and feisty Lieutenant Colonel Donald M. "Buck" Schmuck, 35, University of Colorado 1938, had taken over command of the 1st Battalion, 1st Marines, on 8 November. In World War II he had fought as a company commander at Bougainville and Peleliu, and later served at Okinawa. On the night of 26 November the Chinese probed his perimeter at Chinhung-ni at the foot of Funchilin Pass with a series of light attacks. Patrols sent out by Schmuck the next day failed to make contact. That night the Chinese hit his perimeter with another tantalizing, easily repulsed, light attack. Schmuck sent out more patrols during the next two days. What they found or did not find caused him to conclude that a Chinese battalion that attacked him at night and hid in the houses to his west during the day was pestering him.

Photo courtesy of Sgt Norman L. Strickbine, USA

Parachutes laden with ammunition and supplies blossom as they are dropped by an Air Force C-119 of the Far East Air Force's Combat Cargo Command to the entrapped RCT-31 east of the reservoir. Such aerial delivery had mixed results. Much of what was dropped fell outside American lines and into Chinese hands.

A patrol sent out from Captain Wesley Noren's Company B on the 29th more or less confirmed Schmuck's conclusion. Schmuck decided to attack the suspected Chinese position on the following day, using Captain Robert H. Barrow's Company A and a part of Noren's company, reinforced with 81mm and 4.2-inch mortars. Battery F, 11th Marines, under First Lieutenant Howard A. Blancheri, laid down preparatory 105mm howitzer fire, the infantry swept forward, and, in the words of Major William L. Bates, Jr., the battalion's Weapons Company commander, "ran the Chinese right out of the country." The houses that sheltered the Chinese were burned. There was no more trouble at Chinhung-ni.

Task Force Drysdale Formed

On the evening of 28 November three disparate units—41 Commando, Royal Marines; Company G, 3d Battalion, 1st Marines; and Company B, 31st Infantry, 7th Infantry Division—had crowded into the perimeter at Koto-ri after an uneventful motor march up from the south. Puller pasted the three units together into a task force, giving command to Lieutenant Colonel Douglas B. Drysdale of the Royal Marines, with orders to fight his way through to Hagaru-ri the next day.

Drysdale barely had time to uncrate his newly issued American 81mm mortars and Browning machine guns. He moved out at 0945 on 29 November, his truck-borne column followed by a serial of headquarters troops on its way to the new division headquarters at Hagaru-ri. Drysdale's plan was for his Royal Marines to lead off with an assault against the Chinese entrenched on the right of the road just north of Koto-ri. Captain Carl Sitter's George Company—reinforced with a provisional platoon of water-cooled machine guns, rocket launchers, and 81mm mortars—was to follow with an assault against Hill 1236, a mile-and-a-half north of Koto-ri. The soldiers of Baker Company, 31st Infantry, would be in reserve.

The Royal Marines took their objective without much trouble, but Sitter's Marines ran into seri-

Photo by PFC C. Wehner, Jr., National Archives Photo (USMC) 127-N-A6593

LtCol Donald M. Schmuck received a Silver Star for his command of 1st Battalion, 1st Marines, at Chinhung-ni from 26 November until 11 December. Making the award sometime in late winter 1951, is Col Francis A. McAlister, who had been the division G-4 but who by this time had succeeded Col Puller in command of the 1st Marines.

column to punch a way through. Resignedly, Drysdale resumed his advance at 1350, with 17 tanks leading the way, followed by Sitter's George Company. It was a pulsating advance—short movements followed by pauses while Chinese strong points were reduced with 90mm and machine gun fire. Progress was slow and George Company took heavy losses.

More tanks—Company B, 1st Tank Battalion—arrived at Koto-ri at about 1500. Puller ordered their commander, Captain Bruce F. Williams, to leave one platoon with Sutter's 2d Battalion and to join the rear of the Drysdale column with his remaining two platoons. Meanwhile, Puller had dispatched a platoon from Company E, 2d Battalion, to assist in the evacuation of Drysdale's casualties. The platoon did not get back into the Koto-ri perimeter until about 1600.

ous resistance before taking Hill 1236. The British and American Marines then moved together about a mile farther up the road where they were stopped by Chinese machine gun and mortar fire coming from Hill 1182. Drysdale received a message from Puller telling him that three platoons of tanks would arrive by 1300 to help. The tanks—two platoons of Pershings from Captain Bruce W. Clarke's Company D, 1st Tank Battalion, and the tank platoon with Shermans from the regiment's Anti-Tank Company—had just arrived at Koto-ri at noon. Drysdale ordered Sitter to break off the action and come back down to the road and await the tanks.

Drysdale found Captain Clarke an "opinionated young man." Drysdale wanted the tanks distributed throughout the length of the column. Clarke insisted that they be kept together at the head of the

This sketch by Sgt Schofield shows the meeting at Hagaru-ri of two U.S. Marines with two Royal Marines of 41 Independent Commando. The professionalism and sangfroid of the British Marines impressed their American counterparts, who, in turn, impressed the British with their dogged fighting qualities.

Sketch by Sgt Ralph Schofield, USMCR

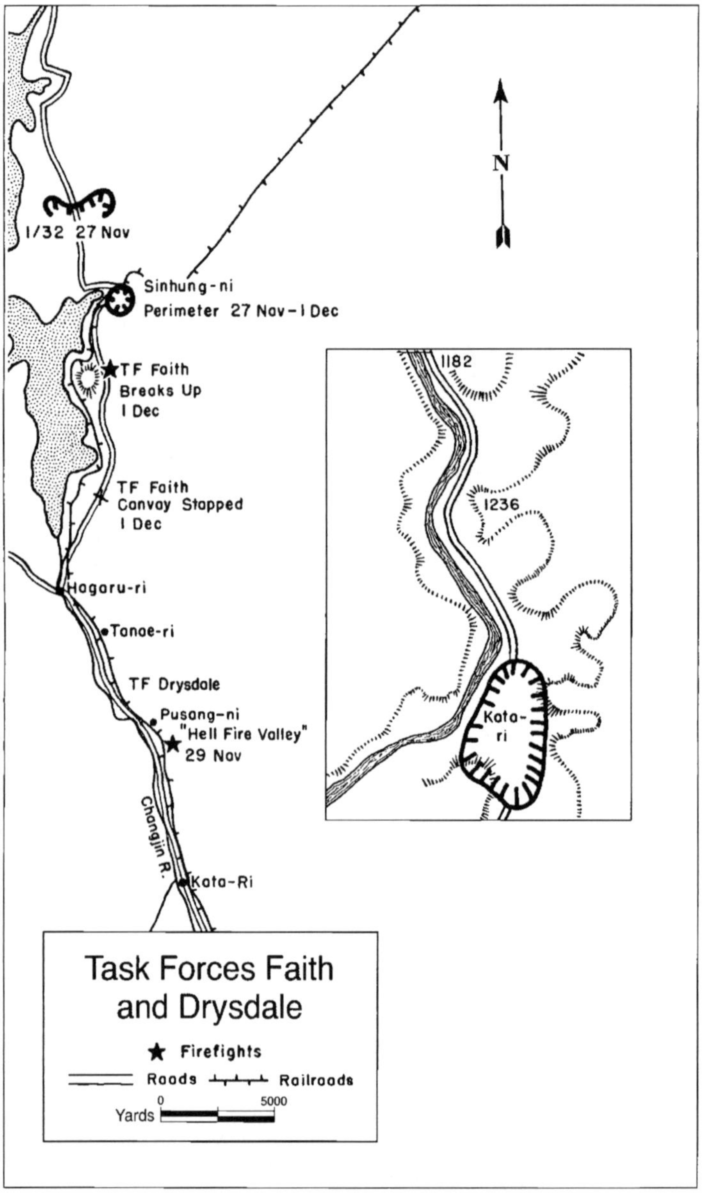

Task Force Drysdale came to a halt about four miles north of Koto-ri at about the same time. Shortly thereafter the Chinese began pounding the northern face of the Koto-ri perimeter with mortar fire followed by a company-sized attack that was easily contained by Easy Company.

Clarke and Williams, the tank commanders, advised Drysdale and Sitter that they thought the tanks could get through to Hagaru-ri but were dubious about further movement by trucks. Drysdale put the decision of a further advance up to division. Smith ordered him to continue. The tanks needed to refuel and this took more time. When the column did plunge forward unit integrity was lost and combat troops became intermingled with headquarters elements.

At the midway point to Hagaru-ri there was a valley, about a mile long, high ground on one side and the Changjin River and more hills on the other—Drysdale would name it "Hell Fire Valley." It became the scene of an all-night fight. The column broke in half. George Company, three-quarters of 41 Commando, and a few soldiers, led by tanks from Company D, continued on toward Hagaru-ri. The remainder of 41 Commando; most of Company B, 31st Infantry; and nearly all other headquarters personnel were left on the road which the Chinese closed behind them. The Chinese chopped away at them. The best protection the stalled half of the convoy could find were the shallow ditches on each side of the road. Lieutenant Colonel Arthur A. Chidester, 37, University of Arkansas 1935, the assistant division G-4 and the senior officer in the group, attempted to turn his truncated column around and return to Koto-ri. He was wounded and captured. His place was taken by Major James K. Eagan, soon also wounded and taken prisoner.

The half-column that had been left behind coalesced into one large perimeter and three small ones strung out over a distance of close to a mile. Farthest north, near the hamlet of Pusong-ni, was the largest perimeter, a hodgepodge of about 140 men including Associated Press photographer Frank "Pappy" Noel. Senior officer was Major John N. McLaughlin, 32, Emory University 1941, an assistant division G-3 and a well-decorated veteran who had fought with the 5th Marines at Guadalcanal,

National Archives Photo (USMC) 127-N-A5354

The village of Koto-ri, midway between Hagaru-ri and the Funchilin Pass, was held by the 2d Battalion, 1st Marines, and was the site of Col "Chesty" Puller's regimental command post. Units moving north and south staged through here. Unseen are the fighting holes of the Marine infantry that encircled the camp.

Cape Gloucester, and Peleliu in World War II.

There was some hope that the Company B tanks from Koto-ri would come to the rescue of the four ragged perimeters, but the tanks were stopped by the Chinese at the defile formed by Hills 1236 and 1182, the same hills captured earlier but now reoccupied by the Chinese. The southernmost group on the road worked its way back into the Koto-ri perimeter by 2200 without much trouble. The middle group, mostly headquarters personnel, also made it back by 0230, losing most of its trucks along the way. Its leader, Lieutenant Colonel Harvey S. Walseth, the division G-1, was wounded. (Lieutenant Colonel Bryghte D. Godbold would take his place on the general staff.) By dawn all of the Company B tanks had returned to Koto-ri.

The troops remaining trapped in Hell Fire Valley and still hoping to be rescued by the tanks knew none of this. The Chinese meanwhile seemed more interested in looting the trucks than annihilating the defenders. Major McLaughlin tried sending patrols back to the south to link up with the other perimeters. They were beaten back. He gathered his wounded in a ditch and prayed for daylight and the arrival of Marine Corps aircraft overhead. By 0200 he was out of grenades. A 75mm recoilless rifle, gallantly manned by U.S. soldiers, was knocked out and all its crew killed or wounded. Associated Press photographer Noel and two men attempted to run the gantlet in a jeep and were captured.

The Chinese at about 0430 sent several prisoners into McLaughlin's position bearing a demand that the Americans surrender. McLaughlin and a British Marine went out under a white flag to parley. In a desperate act of bravado McLaughlin pretended that the Chinese wished to surrender to him, but the enemy was neither impressed nor amused. They gave him 10 minutes to capitulate or face an all-out assault. McLaughlin, with only about 40

A gaggle of Marines, looking like giant penguins in their hooded parkas, watch with great interest an air strike against a hill outside Koto-ri on 28 November. The next day these Marines very probably could have been part of Task Force Drysdale, which would depart Koto-ri to fight its way to Hagaru-ri.

Photo by Cpl W. T. Wolfe, Department of Defense Photo (USMC) A4865

able-bodied defenders and almost no ammunition, reluctantly decided to surrender but with the condition that his most serious wounded be evacuated. The Chinese agreed to his terms. The Chinese did not live up to their promise, but they did, however, permit some of the wounded to be placed in houses along the road where they might eventually be found.

While McLaughlin was negotiating his surrender, some few Americans and British Marines and a considerable number of U.S. soldiers managed to slip away from the smaller perimeters to the south. This group, led largely by Major Henry W. "Pop" Seeley, Jr., 33, Amherst College 1939, made its way successfully back to Koto-ri. Seeley had spent four years in the Pacific during World War II and had been well decorated for his service.

Drysdale had continued his start-and-stop progress with Company D's tanks, Company G, and the larger part of 41 Commando, not knowing what had happened to the rear half of his haphazard command. One of the tanks was knocked out by a satchel charge. Drysdale received a grenade fragment in the arm and deferred command of the column, momentarily, to Sitter.

Well after dark the first of Company D's tanks, leading the column, burst through Hagaru-ri's south roadblock, flattening one of Weapons Company's jeeps in the process. Sitter's George Company came into the perimeter, battered but intact. The Royal Marine Commando, in accordance with its training, split into small groups. For most of the night, U.S. Marines on the perimeter were treated to English accents shouting, "Don't shoot, Yanks. We're coming through." Royal Marine troop commander Lieutenant Peter Thomas said later, "I never thought I should be so glad to see an American." At about midnight, Lieutenant Colonel Drysdale, blood dripping down his arm, gave Lieutenant Colonel Ridge a side-winding salute and reported 41 Commando present for duty.

Maj John N. McLaughlin, left, taken prisoner during the disastrous advance of Task Force Drysdale from Koto-ri to Hagaru-ri, is welcomed back from captivity, still in his Chinese cap and coat, on 5 September 1953, by BGen Joseph C. Burger, then the assistant division commander of the 1st Marine Division.
Photo by TSgt Jack E. Ely, Department of Defense Photo (USMC) A1744606.

By best estimates, Task Force Drysdale had begun the day with 922 officers and men. Something like 400 men reached Hagaru-ri, another 300 hundred found their way back to Koto-ri. Killed in action and missing in action were estimated at 162. Another 159 men were identified as wounded. Forty-four Marines, originally listed as missing, were taken prisoner. Of these, just 25 either escaped or survived their captivity. Chidester and Eagan were among those wounded who died in captivity. The column had started with 141 vehicles and 29 tanks. Of these, 75 vehicles and one tank were lost.

George Company Goes Up East Hill

Lieutenant Colonel Ridge's command group had remarkably good intelligence as to the extent of the enemy outside the Hagaru-ri perimeter, the information often brought in by "line-crossers," plainclothes Korean agents who boldly moved in and out of the perimeter. The Chinese *58th Division's* reported intentions to renew its attack against Hagaru-ri seem to have been thwarted by well-placed air attacks during the day and heavy artillery and mortar fires during the night. Ridge's supporting arms coordinator also experimented with night close air support, using converging bands of machine gun tracer fire to point out targets to the Corsair "night hecklers" overhead.

At 0800 on 30 November, the morning after George Company's arrival and a scant night's sleep, Ridge ordered Sitter to pass through Myers' position on East Hill and continue the attack. Drysdale's 41 Commando was held in reserve. This company-sized force of highly trained Royal Marines gave Ridge a small but potent maneuver element, far more promising than the scratch reserve formations he had been forced to use. Drysdale and his officers spent much of the day reconnoitering possible counterattack routes and acquainting themselves with supporting fire plans. "I felt entirely comfortable fighting alongside the Marines," said Drysdale.

Sitter, stoic and unflappable, sent out his 1st and 2d Platoons to pass through Myers' toehold on the hill. They were then to attack on both sides of the ridge. The 3d Platoon and two platoons of Able Company engineers would follow in reserve. Progress was slow and Sitter used his reserve to envelop the Chinese right flank. The attack bogged down and Sitter asked for permission to set up defensive positions on the ground previously held by Myers who had withdrawn his meager force. Corsairs were brought in and worked over the crest of the hill again and again, but George Company could not take the contested ground.

National Archives Photo (USAF) AC40314

Ammunition and supplies for the embattled Marines descend from a C-119 Flying Boxcar of the Far East Air Forces Combat Cargo Command. This was the only photograph the U.S. Air Force photographer was able to take before 23-degree-below-zero weather froze the shutter on his camera.

That same day at Hagaru-ri, 30 November, Colonel Brower, commanding the 11th Marines, came down with a serious liver infection. Command of the regiment passed to his executive officer, Lieutenant Colonel Youngdale.

Disaster Threatens RCT-31

Sung Shi-lun was amazingly well informed as to exactly what his opponents were doing. Chinese reconnaissance was good; and Korean civilians, including line crossers, were at least as useful to the Chinese as they were to the Americans. Moreover, he apparently had a serviceable quantity of signal intelligence from radio intercepts. Stymied by the Marines' stubborn defense at Yudam-ni and Hagaru-ri, he decided to finish off the U.S. Army forces east of the reservoir by adding the weight of the *81st Division* to the *80th Division* already engaged against Task Force Faith.

The curious command relationships at Yudam-ni continued. Without a common commander in place on the ground, the two collocated regiments pursued their separate missions. Smith had issued an order on the afternoon of 29 November directing Murray to assume responsibility for the protection of the Yudam-ni area with his 5th Marines, while Litzenberg was to employ the entire 7th Marines in clearing the MSR to Hagaru-ri "without delay."

Almond Issues New Orders

At noon on 29 November, Almond departed Haneda airfield in Japan on his return flight from his meeting with MacArthur. Enroute to Yonpo he directed his G-3 and other staff members to commence planning the break-off of the offensive and the consolidation of the corps. When Almond arrived at his war room in Hamhung he saw that, in addition to the predicament of the 1st Marine Division and RCT-31 at the Chosin Reservoir, his remaining forces were in considerable disarray. Soule's 3d Infantry Division was headed in two different directions. A CCF column at Sachang far to the southwest of Yudam-ni had already engaged the division's 7th Infantry. The remaining two regiments, the 15th and 65th, were regrouping at Yonghong on the coast preparatory to attacking west, in accordance with orders to relieve pressure on the Eighth Army's dangling flank. In Barr's 7th Infantry Division, MacLean's RCT-31 was already isolated and heavily engaged east of the reservoir. Barr's remaining regiments, the 17th and 32d Infantry, were pulling back to the Pungsan area. By 2100 that evening X Corps Operation Order Number 8, providing for the discontinuance of the attack to the northwest and the withdrawal of forces into the Hamhung-Hungnam perimeter, was ready for Almond's approval.

By that order, Almond placed under Smith's command all Army troops in the Chosin Reservoir area, including Task Force Faith and elements at Hagaru-ri, effective 0800 the next morning. Along with the assignment of these troops came a highly optimistic order from X Corps to Smith to "redeploy one RCT without delay from Yudam-ni area to Hagaru area, gain contact with elements of the 7th Inf Div E of Chosin Reservoir; coordinate all forces in and N of Hagaru in a perimeter defense based on Hagaru; open and secure Hagaru-Koto-ri MSR."

At 0600, 30 November, Litzenberg and Murray issued a joint order for the breakout. (Smith did admit in his log entry for 30 November: "An ADC [assistant division commander; that is, Craig] would have come in handy at this point.") That same morning, Almond gave the senior members of his staff a fuller briefing on the MacArthur decision to go over to the strategic defensive. He made it known that he had also issued orders to the ROK I Corps to pull back. By this time the 3d ROK Division was at Hapsu and the Capital Division above Chongjin. They were now to withdraw to Songjin, a deepwater port about 100 miles northeast of Hungnam.

General Barr, who had established an advance command post at Hungnam, was among those present. After the briefing, Barr—whether at Almond's suggestion or on his own initiative is not clear—flew to Hagaru-ri. There he met with O.P. Smith and Hodes and then borrowed a Marine helicopter to go forward to Faith's position. Smith asked Hodes to draft a message advising Faith that his command was now attached to the 1st Marine Division. Barr at this point was out of the operational chain-of-command to Task Force Faith, but RCT-31 was still, of course, part of the 7th Infantry Division. Barr told Smith that he was recalling Hodes from the Chosin Reservoir area to avoid any misunderstanding as to command arrangements. (Hodes would pay a last visit to Hagaru-ri on 2 December,)

Barr arrived at Faith's command post shortly before noon. He presumably informed Faith of the changed command status, either in substance or by delivering the Hodes dispatch.

On his return to Hagaru-ri, Barr agreed with Smith that Task Force Faith, with Marine and Navy close air support, could extricate itself and get back to Hagaru-ri. Almond arrived at about this time and met with Smith, Barr, and Hodes at Smith's forward command post, a few hundred yards from where Ridge's Marines were contending for possession of East Hill. Almond announced that he had abandoned any idea of consolidating positions in the Chosin Reservoir area. A withdrawal would be made posthaste to Hungnam. Almond authorized Smith to destroy or burn all equipment that would impede his movement. Resupply would be by air. Smith demurred: "I told him that my movements would be governed by my ability to evacuate the wounded, that I would have to fight my way back and could not afford to discard equipment and that, therefore, I intended to bring out the bulk of my equipment."

Almond shrugged. He then directed Smith and Barr to work out a time-phased plan to pull back the three Army battalions of RCT-31 making up "Task Force Faith." Furthermore, if Faith failed to execute his orders, Almond

National Archives Photo (USA) 111-SC353607

Cases of ammunition are sorted out and lashed together at Yonpo airfield on 29 November before loading into C-47 transports of the Far East Air Force's Combat Cargo Command. This ammo was intended for RCT-31, better known as Task Force Faith, east of the reservoir. First airdrops directly from Japan would be to Hagaru-ri on 1 December.

opined that he should be relieved.

Almond later said that in addition to general instructions on withdrawal of forces and a specific plan for the withdrawal of RCT-31, he also asked Smith for an explicit plan for the evacuation of both Army and Marine wounded by way of the airstrip being completed at Hagaru-ri.

Almond told Barr that on his flight up to Hagaru-ri he had passed over a column of trucks halted on the road a few miles south of Koto-ri and recognized it as the 2d Battalion, 31st Infantry, which was working its way up from Chinhung-ni. He suggested to Barr that he relieve the slow-moving battalion commander, West Pointer Lieutenant Colonel Richard F. Reidy. Barr objected, saying he did not know the situation confronting the battalion. He may also have reminded Almond that all movements up to Koto-ri were being coordinated by X Corps. Almond telephoned (by radio link) his chief of staff, the able Major General Clark L. Ruffner, and ordered him to expedite the movement north of the 2d Battalion, 31st Regiment, to join Puller's forces at Koto-ri. Before leaving, Almond told Barr and Smith that Soule's 3d Infantry Division was doing a "magnificent" job of covering the gap with the Eighth Army.

After Almond had departed, Barr and Smith agreed that not much could be done for RCT-31 until the 5th and 7th Marines arrived at Hagaru-ri from Yudam-ni. Smith did order Litzenberg and Murray to expedite their withdrawal, destroying any supplies and equipment that had to be abandoned.

Ruffner sent Captain Joseph L. Gurfein, West Point 1944, to get Reidy, a short, chunky man whose face showed the marks of his boxing days at the Academy, moving. Gurfein found Reidy and his battalion, which had only two rifle companies, stalled about three miles outside Puller's position. (The other rifle company had gone forward earlier, had been with Task Force Drysdale, and was now in place, more or less intact, at Koto-ri.) Reidy's battalion, urged on by Gurfein, made a faltering night attack and eventually pushed its way into Koto-ri. Puller gave the battalion a sector of Koto-ri's defensive perimeter. He also ordered Reidy (who was suffering from a badly infected foot) to take charge of the sizable number of soldiers from various units—including a detachment of the 185th Engineer Battalion—that had now collected at Koto-ri.

With Task Force Faith

During the daylight hours of 30 November, Don Faith worked out counterattack plans to meet a penetration of his perimeter. The Chinese, not waiting for nightfall, began their attack in the afternoon. Task Force Faith's perimeter at Sinhung-ni was now isolated and alone with no friendly forces between it and Hagaru-ri. By midnight the attack against Faith's perimeter had built up to unprecedented intensity. There were penetrations, but Faith sealed these off with local counterattacks. At the aid station, medical supplies were completely exhausted. The dead, frozen stiff, were laid out in rows stacked about four feet high.

Meanwhile, well to Faith's rear, headquarters elements of RCT-31 and the 31st Tank Company at Hudong-ni, with 1st Marine Division approval, had fallen back to Hagaru-ri. Two disabled tanks had to be abandoned along the four-mile route, but otherwise the march, about 325 soldiers altogether, was made without incident. The regimental S-3, Lieutenant Colonel Berry K. Anderson, the senior Army officer present, was in charge. A new 31st Infantry headquarters was being formed at Hamhung, with Colonel John A. Gavin, USA, as its designated commander, but it was not sent forward to Hagaru-ri.

Intimations of an Evacuation

By the end of November it was increasingly obvious that Rear

Admiral James Doyle, who had landed the Marines at Inchon and again at Wonsan, was now going to have to lift them out of Hungnam as part of a massive amphibious withdrawal. Doyle, as Commander, Task Force 90, issued plans on 28 November for a redeployment of United Nations forces. Doyle's plans called for the division of his Task Force 90 into two amphibious task groups. Task Group Three, under Rear Admiral Lyman A. Thackrey, would provide for amphibious evacuation on the west coast of Eighth Army units if required. Task Group One, under Doyle's immediate command, would execute the amphibious evacuation of east coast ports, primarily Hungnam. Task Group Three, with two-thirds of the amphibious force, would go to the west coast where the situation, at that moment, seemed more critical. There would not be nearly enough amphibious ships for these tasks; there had to be an enormous gathering of merchant shipping. Vice Admiral Joy's deputy chief of staff, newly promoted Rear Admiral Arleigh Burke, largely ran this effort.

The carrier-based aircraft brought together for the Inchon landing and then the Wonsan landing had by mid-November been largely dispersed. Left with Task Force 77 were the fast carriers *Leyte* (CV 32) and *Philippine Sea* (CV 47). Also still on station was the escort carrier *Badoeng Strait* (CVE 116)—"Bing-Ding" to the Marines and sailors—but her sister ship *Sicily* (CVE 118), having dropped off VMF-214, the "Blacksheep" squadron, at Wonsan, was in port in Japan. Major Robert P. Keller had commanded the squadron until 20 November when he was detached to become the Marine air liaison officer with Eighth Army and Fifth Air Force. In World War, Keller served in the Pacific with Marine Fighter Squadrons 212 and 223 and was credited with at least one aerial victory. Command of VMF-214 was taken over by Major William M. Lundin.

The big carrier *Valley Forge* (CV 45), which on 3 July had been the first carrier to launch combat missions against North Korean invaders, was on her way home for a much-needed refit. Now the emergency caused her to turn about and head for the Sea of Japan. Also on the way was the *Princeton* (CV 37), hurriedly yanked out of mothballs. But until these carriers could arrive, tactical air operations, including all important close air support, would have to be carried out by shore-based Marine squadrons and Navy and Marine squadrons in the *Leyte, Philippine Sea,* and *Badoeng Strait.*

On 1 December the Far East Air Forces relinquished control of all tactical air support of X Corps to the 1st Marine Aircraft Wing, which, in the words of air historian Richard P. Hallion, "performed brilliantly." The *Princeton* arrived on station on 5 December. By then Task Force 77 was giving the Chosin exodus its full attention.

Carrier-based Vought F4U Corsairs and Douglas AD Skyraiders were the workhorses of Task Force 77. A typical ordnance load for the Corsair on a close air support mission was 800 rounds for its 20mm guns, eight 5-inch rockets, and two 150-gallon napalm bombs. With this load the plane had an endurance of two-and-one-half hours. The Navy's Skyraider, much admired by the Marines, packed an ordnance load comparable to a World War II Boeing B-17 heavy bomber, commonly 400 rounds of 20mm, three 150-gallon napalm bombs, and either twelve 5-inch rockets or twelve 250-pound fragmentation bombs. The Skyraider could stay in the air for four hours with this load.

Bad Night at Hagaru-ri

The Marines at Hagaru-ri had another bad night on 30 November. The Chinese, the *58th Division* now augmented by the *59th*, assaulted Ridge's weary defenders once again in an attack pattern that repeated that of the night of the 28th. One regiment or more came in against the southwest face of the perimeter and unfortunately for them hit Item Company's well-entrenched position. First Lieutenant Joseph Fisher, always optimistic in his counting of enemy casualties, guessed that he killed as many as 500 to 750 of them. His own losses were two Marines killed, 10 wounded.

Another Chinese regiment came across the contested ground on East Hill hitting the reverse slope defenses held chiefly by Sitter's Company G and 1st Engineer Battalion's Companies A and B. General Smith watched the fight from the doorway of his command post, two-thirds of a mile away. Some ground was lost. Ridge sent up a portion of his precious reserve, 41 Commando, to reinforce Company G, and the lost ground was retaken by early morning.

After the night's action, Ridge, the defense force commander, came to see Smith at about 0900. "He was pretty low and almost incoherent," Smith wrote in his log for 1 December. "The main trouble was loss of sleep. He was much concerned about another attack. He felt with the force available to him he could not hold both the airstrip and the ridge [East Hill] east of the bridge. I told him he would have to hold both and would have to do it with what we had."

Casualty Evacuation and Resupply

Friday, 1 December—although no one thought of it that way at the time—was the turning point of the campaign.

Lieutenant Colonel John Partridge's 1st Engineer Battalion had succeeded in hacking out the semblance of an airstrip from the frozen earth in 12 days of around-the-clock dangerous work, the engineers at times laying down their tools to take up rifles and machine guns. Heroic though the engineering effort was, the airstrip was only 40 percent complete. Its rough runway, 50-feet wide and 2,900-feet long, fell considerably short of the length and condition specified by regulations for operation of transport aircraft at those altitudes and temperatures. Smith decided that the urgency of the evacuation problem was such that the uncompleted airfield must be used, ready or not.

Its impossible load of casualties was overwhelming the division field hospital—a collection of tents and Korean houses. Navy Captain Eugene R. Hering, the division surgeon, met with General Smith that morning. Two additional surgical teams had been flown in by helicopter from Hungnam. The two companies, Charlie and Easy, of the 1st Medical Battalion, already had some 600 patients. Hering expected 500 more casualties from Yudam-ni and 400 from the Army battalions east of the reservoir. Grim as his prediction of casualties was, these estimates would prove to be much too low.

LtCol Murray, exact location and time unknown, but possibly mid-November at Hagaru-ri, briefs Maj Vincent J. Gottschalk, commanding officer of VMO-6, on the locations of units of his regiment. Between 27 November and 1 December, VMO-6's helicopters flew 152 medical evacuation missions in addition to many other reconnaissance and liaison missions.

National Archives Photo (USMC) 127-N-A130916

Until the airstrip was operational, aerial evacuation of the most serious cases had been limited to those that could be flown out by the nine helicopters and 10 light aircraft of Major Gottschalk's VMO-6, which also had many other missions to perform. From 27 November to 1 December, VMO-6, struggling to fly at the cold, thin altitudes, had lifted out 152 casualties—109 from Yudam-ni, 36 from Hagaru-ri, and 7 from Koto-ri. One of Gottschalk's pilots, First Lieutenant Robert A. Longstaff, was killed on an evacuation flight to Toktong Pass.

At 1430 on Friday afternoon, the first Air Force C-47 transport touched down on the frozen snow-covered runway. A half-hour later the plane, loaded with 24 casualties, bumped its way off the rough strip into the air. Three more planes came in that afternoon, taking out about 60 more casualties. "It takes about a half hour to load a plane with litter patients," Smith noted in his log. "Ambulatory patients go very much faster." The last plane in for the day, arriving heavily loaded with ammunition, collapsed its landing gear and had to be destroyed.

Because of Smith's foresight,

Photo by Sgt William R. Keating, Department of Defense Photo (USMC) A5683

Casualty evacuation from Hagaru-ri airstrip began on 1 December, although the condition of the field fell far short of what safety regulations required. Here, on a subsequent day, an ambulance discharges its cargo directly into an Air Force C-47 or Marine Corps R-4D transport. Weapons of other, walking wounded, casualties form a pile in the foreground.

the X Corps' deputy chief of staff, visited Smith. McCaffrey, who had been Almond's chief of staff in the 92d Infantry Division in Italy, outlined for Smith the plan for constriction into a Hungnam perimeter and its subsequent defense. Soule's 3d Infantry Division was to move elements to the foot of Funchilin Pass and provide a covering force through which the 1st Marine Division would withdraw. The 1st Marine Division would then organize a defensive sector west and southwest of Hungam. The 7th Infantry Division would occupy a sector northeast and north of Hungnam.

The consolidation of X Corps in the Hamhung-Hungnam area included the evacuation of Wonsan to the south. Major General Field Harris, commanding the 1st Marine Aircraft Wing, ordered MAG-12—the Marine aircraft group had three tactical squadrons and a headquarters

Hagaru-ri was already stockpiled with six days of rations and two days of ammunition. The first airdrop from Air Force C-119 "Flying Box Cars" flying from Japan was on that same critical 1st of December. The drops, called "Baldwins," delivered prearranged quantities of ammunition, rations, and medical supplies. Some drops were by parachute, some by free fall. The Combat Cargo Command of the Far East Air Forces at first estimated that it could deliver only 70 tons of supply a day, enough perhaps for a regimental combat team, but not a division. By what became a steady stream of transports landing on the strip and air drops elsewhere the Air Force drove its deliveries up to a 100 tons a day.

Toward the end of the day Lieutenant Colonel William J. McCaffrey, West Point 1939 and

Mountains of supplies were rigged at Yonpo Airfield by Capt Hersel D. C. Blasingame's 1st Air Delivery Platoon for parachute drop, free airdrop, or simply as cargo for the transports flying into Hagaru-ri. The 1st Marine Division needed 100 tons of resupply a day, delivered by one means or another.

Photo by SSgt Ed Barnum, Department of Defense Photo (USMC) A130436

squadron—to move up from Wonsan to Yonpo. The group commander, Colonel Boeker C. Batterton, 46, Naval Academy 1928, had spent most of World War II with a naval aviation mission in Peru. MAG-12 completed the movement of its aircraft in one day—that same busy 1 December. Some planes took off from Wonsan, flew a mission, and landed at Yonpo.

East of Chosin

Total strength of RCT-31 has been calculated at a precise 3,155, but of this number probably not more than 2,500 fell under Don Faith's direct command in "Task Force Faith" itself. On the morning of 1 December, Lieutenant Colonel Faith, on his own initiative, began his breakout from Sinhung-ni to the south. He did not have a solid radio link to the 1st Marine Division, and had nothing more than a chancy relay through Marine Captain Edward P. Stamford's tactical air control net. Faith's own 1st Battalion, 32d Infantry, would lead off, followed by the 57th Field Artillery, with the 3d Battalion, 31st Infantry, bringing up the rear. Trucks would be unloaded so as to carry the wounded. The howitzers were spiked. Most jeeps and all inoperable trucks were to be destroyed, as would be all supplies and equipment. In execution the destruction of the surplus was spotty. About 25 to 30 vehicles—all that were still in operating condition—formed up into column. The overriding mission for Task Force Faith was now to protect the truck convoy with its hundreds of wounded. Much reliance would be placed on the automatic weapons fire of the tracked weapons carriers, down to three in number, two quad .50s and one dual 40mm.

The column began to move out at about 1100. The soldiers could see the Chinese in plain sight on the surrounding high ground. Progress was slow. Mortar rounds continued to fall, causing more casualties, and Chinese infantry began pressing in on the column. The fighting for the first half-mile was particularly intense. Officers and noncommissioned officers suffered disproportionate losses. Control broke down.

Captain Stamford and his tactical air control party tried to keep close air support overhead as continuously as possible. Navy Corsairs from the fast carrier *Leyte* came on station at about 1300. With Stamford calling them in, the Corsairs used napalm and rockets and strafed with 20mm cannon. One napalm drop hit close to Faith's command group causing eight or ten casualties. This ghastly accident was demoralizing, but survivors agreed that without close air support the column would never have cleared the perimeter.

Some of the soldiers on the point began to fall back. Faith drew his Colt .45 pistol and turned them around. Panicky KATUSAs—and some Americans—tried to climb into the trucks with the wounded. Riflemen assigned to move along the high ground on the flanks started to drift back to the road. The head of the column reached a blown bridge just north of Hill 1221 at about 1500. Some of the trucks, trying to cross the frozen stream, broke through the ice and had to be abandoned.

The Chinese held the high ground on both sides of a roadblock that now stood in the way and were in particular strength on Hill 1221. Faith, .45 in hand, gathered together enough men to reduce the roadblock. Other small groups of men clawed their way crossways along the slope of Hill 1221. This fight was almost the last gasp of Task Force Faith. In the words of one major, "[After Hill 1221] there was no organization left."

Even so, by dusk the column was within four-and-one-half road miles of Hagaru-ri when a grenade fragment that penetrated his chest just above his heart killed Faith himself. His men propped up his body, with a blanket wrapped around his shoulders, in the cab of a truck—rather like a dead El Cid riding out to his last battle—hoping that word of his death would not spread through the column causing more demoralization. Just what happened to his body after that is not clear.

As the column struggled on southward the Chinese methodically continued their destruction of the convoy, truck by truck. Individual soldiers and small groups began to break away from the column to attempt to cross the frozen reservoir on foot. Task Force Faith, as such, had ceased to exist.

At Hagaru-ri

During the fighting at Hagaru-ri from 28 November through 1 December, Ridge's 3d Battalion, 1st Marines, had suffered 43 killed, 2 missing, and 270 wounded—a total of 315 battle casualties and a third of its beginning strength. The bits and pieces of Marine and Army units that made up the rest of the Hagaru-ri defense force had casualties perhaps this high if not higher. These casualties, however, did not come even close to those suffered by RCT-31 east of the reservoir.

Throughout the night of 1 December survivors of Task Force Faith drifted into the north side of the perimeter at Hagaru-ri, most of them coming across the ice.

Photo courtesy of Sgt Norman L. Strickbine, USA

Moving toward the perceived safety of Hagaru-ri, soldiers of Task Force Faith march in a well dispersed but terribly exposed single column across the snow-covered, frozen surface of Pungnyuri Inlet. The march across the ice would continue throughout the night of 1 December.

During that last night all semblance of unit integrity dissolved. At about 0230 on the morning of 2 December, Marine Captain Stamford appeared in front of Captain Read's artillery battery position. Stamford had been briefly taken prisoner but escaped. By mid-morning, 670 soldier survivors, many of them wounded or badly frostbitten, had found their way into Hagaru-ri warming tents. They had a terrible tale to tell.

Dr. Hering, the division surgeon, reported to Smith that 919 casualties went out on 1 December, but that among them there was a large number of malingerers. "Unfortunately," Smith entered in his log, "there are a good many Army men, not casualties who got on planes. . . . Men got on stretchers, pulled a blanket over themselves and did a little groaning, posing as casualties. . . .Tomorrow we will get this situation under control and will have MPs at the planes. No man will be able to board a plane without a [medically issued] ticket." Smith, who had ordered Army Lieutenant Colonel Berry K. Anderson to organize physically fit soldier survivors into a provisional battalion, now "talked" to Anderson again and told him to get his soldiers under control.

X Corps had set up a clearing station at Yonpo. Triage determined those casualties who would recover in 30 days or less. They went to the 1st Marine Division Hospital in Hungnam, the Army's 121st Evacuation Hospital in Hamhung, or the hospital ship *Consolation* in Hungnam harbor. Casualties expected to require more than 30 days hospitalization were flown on to Japan.

Lieutenant Colonel Olin L. Beall, 52, a quintessential salty old mustang and a great favorite of General Smith, commanded the 1st Motor Transport Battalion, which held a position on the northeast quadrant of the Hagaru-ri perimeter. Smith, not flamboyant himself, liked colorful leaders such as Beall and Puller. Murray, a very different style of leader, said of Beall: "We all agreed that he would have had to lived a thousand years to have done all the things he claimed to have done. But when people began checking up on some of the things that he had said he had done, by God, he had done them."

Beall had enlisted in the Marine Corps on 5 April 1917, the day before war was declared on

Marine wounded await evacuation at Yudam-ni. A total of 109 went out by helicopter. The rest would go out by ambulance or truck once the 5th and 7th Marines broke their way through to Hagaru-ri. Fixed-wing aerial evacuation from Hagaru-ri began on 1 December.

Photo by Sgt Frank C. Kerr, Department of Defense Photo (USMC) A4858

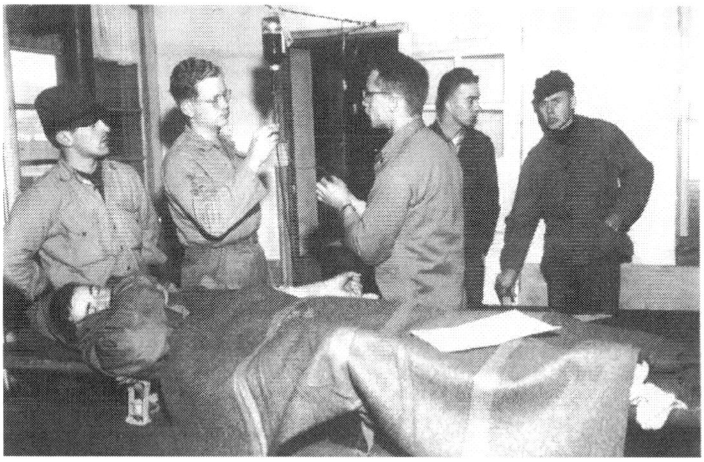

Photo by Cpl Alex Klein, National Archives Photo (USA) 111-SC353559

PFC Ralph C. Stephens, wounded at Koto-ri, receives whole blood at the 1st Marine Division hospital in Hambung on 1 December. Administering the blood are HM 2/C Emmett E. Houston and Lt(jg) Vernon L. Summers. Until the airstrips opened at Hagaru-ri and Koto-ri, evacuation of the wounded was virtually impossible.

Germany. He served in Cuba and Haiti, gained a temporary commission, and went to France just as the war ended. He reenlisted in 1920 and served as an officer—as did many other Marine noncommissioned officers—in the *Gendarmerie d'Haiti* chasing bandits. In 1935 he reached the much-respected warrant grade of Marine gunner. By the end of World War II he was a major and a veteran of Okinawa.

On Saturday, 2 December 1950, Beall led a rescue column of jeeps, trucks, and sleds across the ice looking for other Task Force Faith survivors. Marine Corsairs covered his efforts, flying so low that he said, "I could have scratched a match against their bellies." He brought in 319 soldiers, many of them in a state of shock. The Chinese did little to interfere except for long-range rifle fire.

There is no agreement on exact Army casualty figures. Perhaps 1,050 survivors reached Hagaru-ri. Of these only 385 were found to be physically and mentally fit for combat. These soldiers were given Marine weapons and equipment. Not a single vehicle, artillery piece, mortar, or machine gun of Task Force Faith had been saved. When Almond visited Smith that same Saturday, he had, in Smith's words, "very little to say about the tactical situation. He is no longer urging me to destroy equipment."

Coming Out of Yudam-ni

All day long on Thursday, 30 November, at Yudam-ni the Chinese harried the perimeter with long-range small arms fire and minor probing attacks. As a step in the regroupment of their battered regiments, Litzenberg and Murray organized a provisional battalion made up rather strangely of the combined Companies D and E, 7th Marines; and sections of 81mm mortars from the weapons companies of the 2d and 3d Battalions, 7th Marines; and Companies A and G, 5th Marines. Dog-Easy Company, under First Lieutenant Robert T. Bey, was really no more than two under strength platoons—a Dog Company platoon and an Easy Company platoon Litzenberg gave overall command of this odd assortment to Major Maurice Roach, former commander of the 3d Battalion and now the regimental S-3.

A good part of the reason for forming the battalion was to free Lieutenant Colonel Davis of responsibility for Dog-Easy Company and other attachments. The battalion, which was given its own sector in the perimeter, was assigned the radio sign "Damnation" and that became the short-lived battalion's title. Someone tore up a green parachute to make a neckerchief, the practice caught on, and a green neckerchief became the battalion's badge.

The most difficult task in the disengagement probably fell to Roise's 2d Battalion, 5th Marines, which held a long line stretching from Hill 1426 to 1282. Covered by air and artillery, Roise fell back about a mile from Hill 1426 to Hill 1294. This and other movements freed up Harris' 3d Battalion, 7th Marines, to move to a position astride the MSR about 4,000 yards south of Yudam-ni.

Litzenberg and Murray issued their second joint operation order on the morning of 1 December. Essentially it provided that the 7th Marines would move overland and the 5th Marines would move along the axis of the MSR. Both regiments put what were widely regarded as their best battalions out in front. Davis' 1st Battalion, 7th Marines, would take to the hills; Taplett's 3d Battalion, 5th Marines, down to half-strength, would lead the way down the road. They were to converge in the general vicinity of Fox Company on Toktong Pass. Point for the advance along the road

Photo by Cpl L. B. Snyder, National Archives Photo (USMC) 127-N-A5667

Maj Maurice E. Roach, the 7th Marines jack-of-all-trades, and for the moment commander of a provisional battalion, pauses for a photo with the regimental commander, Col Litzenberg, before leaving Yudam-ni on 1 December. Roach is wearing the Navy parka that was part of the winter uniform, but Litzenberg finds an Army field jacket sufficient for the daytime cold.

would be the single Marine tank that had reached Yudam-ni. Staff Sergeant Russell A. Munsell and another crewman were flown up by helicopter from Hagaru-ri to drive it.

Major William McReynolds' 4th Battalion, 11th Marines, was in general support of both regiments but under the command of neither. The arrangement was made to work with Lieutenant Colonel Harvey A. Feehan, commander of the 1st Battalion, 11th Marines, acting as coordinator of supporting artillery fire. McReynolds' battalion was ordered to shoot up most of its 155mm ammunition before pulling out. Excess gunners were then organized into nine rifle platoons. The guns themselves were to bring up the rear of the convoy.

Roach's "Damnation Battalion" was to have followed Davis across country, but Litzenberg reconsidered and broke up the battalion on 1 December returning its parts to their parent organizations except for Dog-Easy Company, which only had about 100 effectives. Litzenberg passed the orphan company to Murray who, in turn, passed it to Taplett, his advance guard commander.

All available Marine aircraft were to be in the air to cover the withdrawal. They were to be joined by carrier aircraft from Task Force 77. On the ground only the drivers and the critically wounded would move by vehicle; the rest would walk. It was decided to leave the dead at Yudam-ni and a field burial was held for 85 Marines.

The grand parade began at 0800 on 1 December. Taplett's 3d Battalion, 5th Marines, came down from its positions north of Yudam-ni, followed an hour-and-a-half later by Stevens' 1st Battalion. Company B of Stevens' battalion under First Lieutenant John R. Hancock made up the rear guard coming out of the town. Meanwhile the 3d Battalion, 7th Marines, under Harris proceeded to clear both sides of the road leading south, Company H going up Hill 1419 east of the road while the rest of the battalion went against Hill 1542 on the west side. Roise's 2d Battalion, 5th Marines, having cleared the town, relieved Davis' 1st Battalion, 7th Marines, on Hill 1276, freeing up Davis to pursue his overland mission. Harris was slow in taking Hill 1542. By mid-afternoon, an impatient Taplett was in position behind Harris, ready to attack south astride the road even with his right flank somewhat exposed.

Company H, 7th Marines, commanded by First Lieutenant

Photo by Sgt Frank C. Kerr, National Archives Photo (USMC) 127-N-A4862

Marines depart Yudam-ni on 1 December. The general plan for the breakout was that the 5th Marines would follow the axis of the MSR and the 7th Marines would move overland. MajGen Smith did not formally designate an overall commander, but LtCol Murray deferred to Col Litzenberg because of seniority.

The 7th Marines command group breaks camp at Yudam-ni on 1 December in preparation for the march back to Hagaru-ri. The jeep in the foreground carries a radio. Note the long antenna. Below zero temperatures caused problems with battery life. Extra five-gallon cans of gasoline are lashed to the front bumper.

Department of Defense Photo (USMC) A5666

Howard H. Harris, met trouble on Hill 1419. Harris' battalion commander, Lieutenant Colonel Harris, was fully occupied with problems on Hill 1542. Litzenberg, realizing that Hill 1419 was too far from Hill 1542 for a mutually supporting attack, detached How Company from the 3d Battalion and assigned it to Davis' 1st Battalion. Davis, Hill 1419 now his responsibility, sent his Able Company to add its weight to How Company's effort.

(Continued on page 84)

OVERLEAF: *"Band of Brothers" by Colonel Charles H. Waterhouse, USMCR (Ret), widely regarded by many Marine veterans of the Korean War as Waterhouse's masterpiece, shows the column of Marines winding its way down Funchilin Pass.* Courtesy of Col Waterhouse and the Chosin Few Association.

National Archives Photo (USMC) 127-N-A5670

The heaviest guns in the defense of Yudam-ni were the 155mm howitzers of the 4th Battalion, 11th Marines, commanded by Maj William McReynolds. Shown here, the big guns are preparing to leave the perimeter, towed by their tractor prime movers. Excess gunners were reorganized into provisional rifle units.

Meanwhile, Lieutenant Harris had been wounded and Second Lieutenant Minard P. Newton, Jr., had taken over the company. Able Company, under Captain David W. Banks, passed through How Company and took the hill at about 1930.

Davis now stripped his battalion down for its cross-country trek. Everything needed for the march would have to be hand carried. He decided to go very light, taking only two 81mm mortars and six heavy machine guns (with double crews) as supporting weapons from his Weapons Company. His vehicles, left behind with his sick, walking wounded, and frostbite cases as drivers, were to join the regimental train on the road. Baker Company, now commanded by First Lieutenant Joseph R. Kurcaba, led off the line of march, followed by Davis and his command group, then Able Company, Charlie Company, battalion headquarters, and How Company, still attached.

It was a very dark night. The guide stars soon disappeared. The snow-covered rock masses all looked alike. The point had to break trail, through snow knee-deep in places. The path, once beaten, became icy and treacherous. Marines stumbled and fell. Radios would not work reliably. Davis, moving ahead and floundering in the snow, lost touch with the forward elements of his battalion for a time. He continued forward until he reached the point. His map, hurriedly read by a flashlight held under a poncho, told him that they were climbing Hill 1520, the slopes of which were held by the Chinese. Baker and Charlie Companies converged on the Chinese who were about a platoon in strength, taking them by surprise. Davis stopped on the eastern slope of Hill 1520 to reorganize. Enemy resistance had slackened to small arms fire from ridges across the valley but Davis' men were numb with cold and exhausted. At 0300 he again halted his advance to give his Marines a rest, sending out small patrols for security. Now, for the first time, he gained radio contact with regiment.

On the MSR with the 5th Marines

Taplett, meanwhile, was marching southward astride the MSR, led

Before leaving Yudam-ni, as a matter of professional pride, the 5th and 7th Marines "policed up the area." Trash, including remnants of rations, was collected into piles and burned. Slim pickings were left for Chinese scavengers who, desperate for food, clothing, and shelter, followed close behind the departing Marines.

Photo by Sgt Frank C. Kerr, Department of Defense Photo (USMC) A4848

National Archives Photo (USMC) 127-N-A5666

In coming back from Yudam-ni, the 1st Marine Division's large number of road-bound wheeled vehicles was both an advantage and a handicap. They carried the wherewithal to live and fight; they also slowed the march and were a temptation for attack by the Chinese. Here an 11th Marines howitzer can be seen in firing position to cover the column's rear.

by the solitary Pershing tank, followed by a platoon from his How Company and a platoon of ever-useful engineers. His radio call sign, "Darkhorse," suited his own dark visage. He advanced for about a mile before being halted by heavy fire coming from both sides of the road. He fanned out How and Item Companies and they cleared the opposition by 1930.

Taplett gave his battalion a brief rest and then resumed the advance. Item Company, led by Captain Harold Schrier, ran into stiff resistance on the reverse slope of still-troublesome Hill 1520 east of the road. Schrier received permission to fall back to his jump-off position so as to better protect the MSR. The Chinese hit with mortars and an infantry attack. Schrier was wounded for a second time and Second Lieutenant Willard S. Peterson took over the company. Taplett moved George Company and his attached engineers into defensive positions behind Item Company. It was an all-night fight. In the morning, 2 December, 342 enemy dead were counted in front of Item Company. Peterson had only 20 Marines still on their feet when George Company passed through his position to continue the attack against Hill 1520. George and How Company were both down to two-platoon strengths.

As a reserve Taplett had Dog-Easy Company, 7th Marines, detached from the now-dissolved "Damnation Battalion." Dog-Easy Company moved onto the road between How and George Companies. By noon George Company, commanded by Captain Chester R. Hermanson, had taken Hill 1520 and Dog-Easy had run into its own fight on the road. Second Lieutenant Edward H. Seeburger, lone surviving officer of Dog Company, was severely wounded while giving a fire command to the solitary tank. He refused evacuation. (Seeburger faced long hospitalization and after a year was physically retired as a first lieutenant. In 1995 he received a belated Navy Cross.)

Corsairs reduced the roadblock that held up Dog-Easy Company. "Darkhorse" trudged on, How and George Companies on both sides of the MSR and Dog-Easy moving down the middle, followed by the engineers and the solitary tank.

After leaving Yudam-ni, a unit of the 7th Marines, possibly a company of LtCol Davis' 1st Battalion, leaves the road to climb into the hills. Davis' objective was to come down on Toktong Pass from higher ground so as to relieve Capt Barber's embattled Fox Company.

Photo by Sgt Frank C. Kerr, National Archives Photo (USMC) 127-N-A4849

With Davis' Battalion in the Hills

East of the MSR at daybreak, Davis reoriented the direction of his march. The 1st Battalion, 7th Marines, passed over the east slope of Hill 1520 and attacked toward Hill 1653, a mountain a mile-and-a-half north of Toktong Pass. Davis' radios could not reach Barber on Fox Hill nor could he talk directly to the Corsairs circulating overhead. Fortunately opposition was light except for Chinese nibbling against the rear of his column where Company H, 3d Battalion, was bringing up the wounded on litters. Davis converged on Hill 1653 with his three organic rifle companies.

At last, radio contact was made with Captain Barber on Fox Hill. Barber jauntily offered to send out a patrol to guide Davis into his position. Davis declined the offer but did welcome the control of VMF-312's Corsairs by Barber's forward air controller. Just before noon, lead elements of Company B reached Barber's beleaguered position.

Company A halted on the north side of Hill 1653 to provide manpower to evacuate casualties. Twenty-two wounded had to be carried by litter to safety. The regimental surgeon, Navy Lieutenant Peter E. Arioli, was killed by a Chinese sniper's bullet while supervising the task. Two Marines, who had cracked mentally and who were restrained in improvised strait jackets, died of exposure before they could be evacuated. Marines of Kurcaba's Company B celebrated their arrival on Fox Hill with a noontime meal of airdropped rations. They then went on to take the high ground that dominated the loop in the road where the MSR passed through Toktong Pass. First Lieutenant Eugenous M. Hovatter's Company A followed them and the two companies set up a perimeter for the night. Meanwhile the balance of Davis' battalion had joined Barber on Fox Hill. Barber's Company F had suffered 118 casualties—26 killed, 3 missing, and 89 wounded—almost exactly half of his original complement of 240. Six of the seven officers, including Barber himself, were among the wounded.

5th Marines on the Road

At the rear of the column on the MSR, Lieutenant Colonel Roise's 2d Battalion, 5th Marines, the designated rear guard, had troubles of its own on Hill 1276 during the early morning hours of 2 December. Captain Uel D. Peters' Company F was hit hard. Night fighters from VMF(N)-542 came on station and were vectored to the target by white phosphorus rounds delivered by Company F's 60mm mortars. Strafing and rockets from the night fighters dampened the Chinese attack, but the fight continued on into mid-morning with Fox Company trying to regain lost ground. By then it was time for Roise to give up his position on Hill 1276 and continue the march south.

Lieutenant Colonel Jack Stevens' 1st Battalion, 5th Marines, had its fight that night of 1-2 December east of the road, being hit by a Chinese force that apparently had crossed the ice of the reservoir. Stevens guessed the number of Chinese killed at 200, at least 50 of them cut down in front of Charlie Company by machine guns.

Lieutenant Colonel William Harris' 3d Battalion, 7th Marines, meanwhile was continuing to have trouble on Hill 1542. Litzenberg reinforced Harris with a composite unit, called "Jig Company," made up of about 100 cannoneers, headquarters troops, and other individuals. Command of this assortment was given to First Lieutenant Alfred I. Thomas. Chinese records captured later indicated that they thought they had killed 100 Americans in this action; actual Marine losses were something between 30 and 40 killed and wounded.

Yudam-ni to Hagaru-ri, 2-3 December

At the head of the column, Taplett's Darkhorse battalion on the morning of 2 December had to fight for nearly every foot of the way. George Company still had Hill 1520 to cross. Dog-Easy Company was moving along the road itself. South of Hill 1520 at a sharp bend in the road a bridge over a ravine had been blown, and the Chinese covering the break stopped Dog-Easy Company with machine gun fire. Twelve Corsairs came overhead and ripped into the ravine with strafing fire and rockets. Dog-Easy Company, helped by How Company, resumed its advance. The attached engineer platoon, now commanded by Technical Sergeant Edwin L. Knox, patched up the bridge so vehicles could pass. The engineers had started out with 48 men; they were now down to 17. Taplett continued his advance through the night until by 0200, 3 December, he was only 1,000 yards short of Fox Hill. Taplett could only guess where Davis might be with the 1st Battalion, 7th Marines.

To the rear, the Chinese pecked away at the Marines withdrawing from Hills 1276 and 1542. Stevens' 1st Battalion, 5th Marines, continued to provide close-in flank protection. Marine air held off much of the harassment, but the column of vehicles on the road moved slowly and the jeep and truck drivers became targets for Chinese

Photo by Sgt Frank C. Kerr, National Archives Photo (USMC) 127-N-A4840

Dead Marines at a Yudam-ni aid station wait mutely on their stretchers for loading onto the truck that will take them to eventual burial. Once on the march, practice was to carry those who died in the savage fighting in the hills down to the road where they could be picked up by the regiments' Graves Registration detachments.

snipers. That night the Chinese got through to Lieutenant Colonel Feehan's 1st Battalion, 11th Marines, and the artillerymen had to repulse them with howitzer fire over open sights.

Six inches of new snow fell during the night. In the morning, 3 December, Taplett combined the remnants of Dog-Easy Company with George Company and returned the command to First Lieutenant Charles D. Mize, who had had George Company until 17 November. From up on Fox Hill, Davis made a converging attack against the Chinese still holding a spur blocking the way to Hagaru-ri. He pushed the Chinese into the guns of Taplett's battalion. An estimated battalion of Chinese was slaughtered. By 1300, Davis' "Ridgerunners" had joined up with Taplett's "Darkhorses."

Davis and Taplett conferred. The senior Davis now took the lead on the MSR with his battalion. The lone tank still provided the point. The truck column reached Toktong Pass. The critically wounded were loaded onto the already over-burdened vehicles. Less severely wounded would have to walk. Stevens' 1st Battalion, 5th Marines, followed Davis' battalion, passing through Taplett's battalion. Taplett stayed in Toktong Pass until after midnight. Coming up from the rear on the MSR was Roise's 2d Battalion, 5th Marines, followed by Harris' 3d Battalion, 7th Marines, now the rear guard.

Sergeant Robert B. Gault, leader of the 7th Marines Graves Registration Section, came out of Yudam-ni in the column on the MSR with his five-man section and a truck with which to pick up Marine dead encountered along the way. As he remembered it a few months later: "That was the time when there was no outfit, you was with nobody, you was a Marine, you were fighting with everybody. There was no more 5th or 7th; you were just one outfit, just fighting to get the hell out of there, if you could."

Column Reaches Hagaru-ri

The six fighter-bomber squadrons of Field Harris' 1st Marine Aircraft Wing flew 145 sorties on Sunday, 3 December, most of them in close support of the 5th and 7th Marines. Under this aerial umbrella, Davis' 1st Battalion, 7th Marines, marched along almost unimpeded. In the early evening, Ridge sent out Drysdale with 41 Commando, supported by tanks from Drake's 31st Tank Company, to open the door to the Hagaru-ri perimeter. At about 1900, a few hundred yards out, Davis formed up his battalion into a route column and they marched into the perimeter, singing *The Marines' Hymn*. Hagaru's defenders greeted the marchers with a tumultuous welcome. A field mess offered an unending supply of hot cakes, syrup, and coffee. Litzenberg's 7th Marines command group arrived shortly after Davis' battalion and was welcomed into the motor transport area by Litzenberg's old friend, Olin Beall.

In Tokyo that Sunday, MacArthur sent a message to the Joint Chiefs of Staff that X Corps was being withdrawn to Hungnam as rapidly as possible. He stated that there was no possibility of uniting it with Eighth Army in a line across the peninsula. Such a line, he said, would have to be 150 miles long and held alone by the seven American divisions, the combat effectiveness of the South Korean army now being negligible.

Sketch by Sgt Ralph Schofield, USMCR

The 5th and 7th Marines on arrival at Hagaru-ri combat base found hot chow waiting for them. The mess tents, operating on a 24-hour basis, provided an almost unvarying but inexhaustible menu of hot cakes, syrup, and coffee. After a few days rest and reorganization, the march to the south resumed.

The Chinese made no serious objection to the last leg of the march from Yudam-ni to Hagaru-ri until about 0200 on Monday morning, 4 December, when the prime movers hauling eight of McReynolds' 155mm howitzers at the rear of the column ran out of diesel fuel. That halted the column and brought on a Chinese attack. Taplett's battalion, unaware of the break, continued to advance. The artillerymen—assisted by bits and pieces of the 1st and 2d Battalions, 5th Marines, who were on the high ground to the flanks—defended themselves until Taplett could face around and come to their rescue.

It was a bad scene. The eight heavy howitzers had been pushed off the road, perhaps prematurely, and would have to be destroyed the next day by air strikes. A half-mile farther down the MSR was a cache of air-delivered diesel fuel that would have fueled the prime movers. By 0830 the road was again open. Chinese losses were guessed at 150 dead.

At 1400 on Monday, the rear guard, still provided by the 3d Battalion, 7th Marines, marched into Hagaru-ri and the four-day, 14-mile, breakout from Yudam-ni was over. The Marines had brought in about 1,500 casualties, some 1,000 of them caused by the Chinese, the rest by the cold. Smith observed in his log: "The men of the regiments are. . . pretty well beaten down. We made room for them in tents where they could get warm. Also they were given hot chow. However, in view of their condition, the day after tomorrow [6 December] appears to be the earliest date we can start out for Koto-ri."

Reorganization at Hagaru-ri

Ridge's Marine defenders of Hagaru-ri breathed much more easily after the arrival in their perimeter of the 5th and 7th RCTs. A sanguine corporal opined to his company commander: "Now that the 5th and 7th Marines are here, we can be resupplied by air, hold until spring, and then attack again to the north."

General Almond flew into Hagaru-ri on Monday afternoon to be briefed on the breakout plan and while there pinned Army Distinguished Service Crosses on the parkas of Smith, Litzenberg, Murray, and Beall. Almond then flew to Koto-ri where he decorated Puller and Reidy (who had been slow in getting his battalion to Koto-ri) with Distinguished Service Crosses. Nine others, including

Gurfein, who had nudged Reidy into moving, received Silver Stars. Reidy was relieved of his command not much later.

For the breakout, Murray's RCT-5, with Ridge's 3d Battalion, 1st Marines, and 41 Commando attached, would briefly take over the defense of Hagaru-ri while Litzenberg's RCT-7, beginning at first light on Wednesday, 6 December, would march to the south. Puller's RCT-1 would continue to hold Koto-ri and Chinhung-ni. All personnel except drivers, radio operators, and casualties were to move on foot. Specially detailed Marines were to provide close-in security to the road-bound vehicles. Any that broke down were to be pushed to the side of the road and destroyed. Troops were to carry two-days of C rations and one unit of fire, which translated for most into full cartridge belts and an extra bandoleer of ammunition for their M-1 rifles. Another unit of fire was to be carried on organic vehicles. The vehicles were divided into two division trains. Lieutenant Colonel Banks, commanding officer of the 1st Service Battalion, was put in command of Train No. 1, subordinate to RCT-7. Train No. 2, subordinate to RCT-5, was given to Lieutenant Colonel Harry T. Milne, the commander of the 1st Tank Battalion. Although Smith had stated that he would come out with all his supplies and equipment, more realistically a destruction plan, decreeing the disposal of any excess supplies and equipment, was put into effect on 4 December. Bonfires were built. Ironically, loose rounds and canned foods in the fires exploded, causing some casualties to Marines who crowded close to the fires for warmth.

Air Force and Marine transports had flown out over 900 casualties on Saturday, 2 December, from Hagaru-ri, and more than 700 the next day. To the south that Sunday, 47 casualties were taken out by light aircraft from the strip at Koto-ri. But casualties kept piling up and by the morning of Tuesday, 5 December, some 1,400 casualties—Army and Marine—still remained at Hagaru-ri. In a magnificent effort, they were all flown out that day. Altogether, in the first five days of December, by best count, 4,312 men—3,150 Marines, 1,137 soldiers, and 25 Royal Marines—were air-evacuated.

Even a four-engine Navy R5D ventured a landing. Takeoff with a load of wounded in an R5D was so hairy that it was not tried again. An R4D—the Marine equivalent of the sturdy C-47—wiped out its landing gear in landing. An Air Force C-47 lost power on take-off and crashed-landed outside the Marine lines. Marines rushed to

Members of this patrol, moving along the abandoned narrow gauge railroad track that paralleled the main supply route, help along a wounded or exhausted comrade while the point and rear riflemen provide watchful cover. No wounded Marine need worry about being left behind.

Sketch by Sgt Ralph Schofield, USMCR

National Archives Photo (USMC) 127-N-A4909

Aerial evacuation of wounded and severely frostbitten Marines and soldiers from Hagaru-ri saved many lives. During the first five days of December, 4,312 men—3,150 Marines, 1,137 soldiers, and 25 Royal Marines—were air evacuated by Air Force and Marine transports that also brought in supplies and replacements.

the rescue. The plane had to be abandoned and destroyed, but there were no personnel casualties during the entire evacuation process.

During those same first five days of December, 537 replacements, the majority of them recovering wounded from hospitals in Japan, arrived by air at Hagaru-ri. Most rejoined their original units. A platoon sergeant in Weapons Company, 3d Battalion, 1st Marines, wounded in the fighting in Seoul, assured his company commander that he was glad to be back.

Visitors who could wangle spaces on board the incoming transports began arriving at Hagaru-ri. Marguerite "Maggie" Higgins of the *New York Herald-Tribune*, well known to the Marines from both the Pusan Perimeter and the Inchon-Seoul campaigns, was among the gaggle of war correspondents that arrived on Tuesday, 5 December, including former Marine combat correspondent Keyes Beach. Higgins announced her intention to march out with the Marines. General Smith disabused her of her intention and ordered that she be out of the perimeter by air by nightfall.

A British reporter made the impolite error of referring to the withdrawal as a "retreat." Smith patiently corrected him, pointing out that when surrounded there was no retreat, only an attack in a new direction. The press improved Smith's remark into: "Retreat, hell, we're just attacking in a new direction." The new television technology was demonstrated by scenes taken of the aerial evacuation of the casualties and an interview with General Smith and Lieutenant Colonel Murray.

Major General William H. Tunner, USAF, commander of the Combat Cargo Combat and greatly admired by the Marines because of the sterling performance of his command, was one of the visitors. Tunner had flown the Hump from Burma into China during World War II and later commanded much of the Berlin Airlift. He solicitously offered to evacuate the rest of the troops now in Hagaru-ri. Smith stiffly told him that no man who was able-bodied would be evacuated. "He seemed somewhat surprised," wrote Smith.

Almond met with Major General Soule, commander of the 3d Infantry Division, that Tuesday, 5 December, and ordered him to form a task force under a general officer "to prepare the route of withdrawal [of the 1st Marine Division] if obstructed by explosives or whatnot, especially at the bridge site." The site in question lay in Funchilin Pass. Almond apparently did not know that the bridge had already been destroyed. The downed span threatened to block the Marines' withdrawal. Soule gave command of what was designated as "Task Force Dog" to his assistant division commander, Brigadier General Armistead D. "Red" Mead, a hard-driving West Pointer who had been G-3 of the Ninth Army in the European Theater in World War II.

Hagaru-ri to Koto-ri, 6 December

At noon on Tuesday, 5 December, Murray relieved Ridge of his responsibility as Hagaru-ri defense commander, and the battalions of the 5th Marines plumped up the thin lines held by the 3d Battalion, 1st Marines. The Chinese did not choose to test the strengthened defenses, but at about 2000 that evening an Air Force B-26 mistakenly dropped a stick of six 500-pound bombs close to Ridge's command tent. His forward air controller could not talk to the Air Force pilot because of crystal differences in their radios, but an obliging Marine night-fighter from Lieutenant Colonel Max J. Volcansek, Jr.'s VMF(N)-542 came overhead and promised to shoot down any Air Force bomber that might return to repeat the outrage.

The well-liked Max Volcansek, 36, born in Minnesota, had come

Photo by Cpl Arthur Curtis, National Archives Photo (USA) 111-SC354492

After her adventures at Chosin, including her aborted efforts to march out with the Marines, Marguerite "Maggie" Higgins, wearing a Navy parka and shoe-pacs, arrived safely at Haneda Air Force Base, Tokyo, on 15 December. Her broadly smiling traveling companion is Army MajGen William F. Marquat of MacArthur's staff.

into the Marine Corps as an aviation cadet in 1936 after graduating from Macalester College. During World War II he had flown Corsairs while commanding VMF-222 in the Pacific and had scored at least one Japanese plane. During the battle for Seoul he had been wounded and shot down but quickly recovered and continued in command of VMF(N)-542.

The plan of attack for Wednesday, 6 December, called for the 5th Marines to clean up East Hill while the 7th Marines moved south along the MSR toward Koto-ri. Close air support for the attack against East Hill was to be on station at 0700. With a touch of condescension, Murray's Marines told Ridge's Marines to stand back and watch for a demonstration of how a hill should be taken.

Smith wanted to march out with his men, but Shepherd ordered him to fly to Koto-ri. Death or wounding of Smith, or worse, his capture by the Chinese, could not be risked. By this time the lurking presence of seven CCF divisions had been identified by prisoner of war interrogations—the *58th*, *59th*, *60th*, *76th*, *79th*, *80th*, and *89th*. Two more divisions—the *77th* and *78th*—were reported in the area but not yet confirmed.

Later it would be learned that the *26th CCF Army*—consisting of the *76th*, *77th*, and *78th Divisions*, reinforced by the *88th Division* from the *30th CCF Army*, had moved down from the north into positions on the east side of the MSR between Hagaru-ri and Koto-ri. They had relieved the *60th Division*, which had moved into positions south of Koto-ri. Elements of the *60th Division* were preparing for the defense of Funchilin Pass including positions on the dominant terrain feature, Hill 1081. Even farther south the *89th Division* was positioning itself to move against the defenders of Chinhung-ni.

Murray has given a characteristically laconic account of the attack by the 5th Marines against East Hill:

> I had been ordered to take a little hill, and I had Hal Roise do that job. When he got over there, he found about 200 Chinese in a mass, and he captured the whole crowd of them. So we had about 200 prisoners we had to take care of. . . . A lot of them were in such bad shape that we left them there, left some medical supplies, and left them there for the Chinese to come along and take care of them after we left.

It was not quite that simple. Heavy air, artillery, and mortar preparation began at 0700 on Thursday, 6 December. Captain Samuel S. Smith's Dog Company jumped off in the assault at 0900, beginning a fight that would go on until daylight the next morning.

Photo by Sgt Frank C. Kerr, National Archives Photo (USMC) 127-N-A5520

In a ceremony at Masan on 21 December, LtCol Harold S. Roise, the stalwart commander of 2d Battalion, 5th Marines, receives a Silver Star for actions incident to the seizure of Kimpo Airfield after the Inchon landing. Further awards for Roise for heroism at Chosin would come later.

All three rifle companies of Roise's 2d Battalion and Charlie Company of the 1st Battalion were drawn into it. Estimates of enemy killed ran as high as 800 to 1,000. East Hill was never completely taken, but the Chinese were pushed back far enough to prevent them from interfering with the exit of the division from Hagaru-ri.

RCT-7 Attacks South

General Smith planned to close his command post at Hagaru-ri on Wednesday morning, 6 December. Before he could leave General Barr, commander of the 7th Infantry Division, who arrived to check on the status of his soldiers, visited him. The survivors of Task Force Faith coupled with units that had been at Hagaru-ri and Hudong-ni added up to a provisional battalion of 490 able-bodied men under command of Army Lieutenant Colonel Anderson. As organized by Anderson, the "battalion" actually was two very small battalions (3d Battalion, 31st Infantry, under Major Carl Witte and 1st Battalion, 32d Infantry, under Major Robert E. Jones) each with three very small rifle companies. Smith attached Anderson's force to RCT-7 and it was sometimes called "31/7."

Litzenberg had about 2,200 men—about half his original strength—for the breakout to Koto-ri. His attack order put Lockwood's 2d Battalion, with tanks, on the MSR as the advance guard; Davis' 1st Battalion on the right of Changjin River and the MSR; Anderson's provisional Army battalion on the left of the road; and Harris' 3d Battalion on the road as the rear guard.

Lockwood, it will be recalled, had stayed at Hagaru-ri with his command group and much of his Weapons Company while Companies D and E went forward to Yudam-ni and Company F held Toktong Pass. At 0630, tanks from Company D, 1st Tank Battalion, led Lockwood's reunited, but pitifully shrunken battalion out of the perimeter through the south roadblock. Almost immediately it ran into trouble from Chinese on the left side of the road. The morning fog burned off and air was called in. A showy air attack was delivered against the tent camp south of the perimeter, abandoned days earlier by the Army engineers and now periodically infested with Chinese seeking warmth and supplies. Lockwood's two rifle companies—Fox Company and Dog-Easy Company—pushed through and the advance resumed at noon. Meanwhile, barely a mile out of Hagaru-ri, Captain John F. Morris' Company C, 1st Battalion, surprised an enemy platoon on the high ground to the southeast of the hamlet of Tonae-ri and killed most of them.

At 1400 Smith received a reassuring message from Litzenberg that the march south was going well. Smith decided that it was time to move his command post to Koto-ri. His aide, Major Martin J. "Stormy" Sexton, World War II raider, asked the commander of Weapons Company, 3d Battalion, 1st Marines, for the loan of a jeep to take his boss to the airstrip. A 10-minute helicopter ride took Smith and Sexton to Koto-ri where Puller was waiting. Smith began planning for the next step in the withdrawal.

Photo by Sgt Frank C. Kerr, Department of Defense Photo (USMC) A5464

Col Litzenberg's 7th Marines led off the march south from Hagaru-ri on 6 December. Here one of his units pauses at the roadblock held by Weapons Company, 3d Battalion, 1st Marines, before exiting the town to watch a drop of napalm by a Marine Corsair against a camp abandoned by Army engineers, now infested with Chinese.

Meanwhile Lockwood's 2d Battalion, 7th Marines, had run into more serious trouble another mile down the road. Davis' 1st Battalion, up in the hills, could see the enemy; Lockwood's battalion, on the road itself, could not. Fox Company, with some help from Dog-Easy Company and the Army mini-battalions under Anderson, pushed through at about 1500. Davis' battalion continued to play company-sized hopscotch from hilltop to hilltop on the right of the road. By dark, lead elements of RCT-7 were about three miles south of Hagaru-ri. Enemy resistance stiffened and air reconnaissance spoke of Chinese columns coming in from the east, but Litzenberg decided to push on. After two more miles of advance, Lockwood's battalion was stopped in what Drysdale had called Hell Fire Valley by what seemed to be a solitary Chinese machine gun firing from the left. An Army tank solved that problem. Another half-mile down the road a blown bridge halted the column. The engineers did their job, the march resumed, but then there was another blown bridge. At dawn on Thursday things got better. Air came overhead, and 2d Battalion, 7th Marines, had no more trouble as it marched the last few miles into Koto-ri. Through all of this Lockwood, sick with severe bronchitis, had sat numbly in his jeep. Early that morning his executive officer, Major Sawyer, had been wounded in the leg by a mortar

Coming out of Hagaru-ri, Col Litzenberg used LtCol Lockwood's 2d Battalion, 7th Marines, reinforced with tanks, as his advance guard. Here a heavy machine gun squad rests by the side of the road while a M-26 Pershing medium tank trundles by. The M-26 mounted a powerful flat-trajectory 90mm gun.

Photo by Sgt Frank C. Kerr, National Archives Photo (USMC) 127-N-A5469

Photo by Sgt Frank C. Kerr, Department of Defense Photo (USMC) A5428

Marching, encumbered with weapons, packs, and winter clothing, on the road from Hagaru-ri to Koto-ri was heavy work, even without Chinese interference. This unit of the 7th Marines, taking advantage of a halt, manages a quick nap. Most of the fighting was on the high ground on both sides of the MSR, but occasionally the Chinese reached the road.

fragment and was out of action. Major James F. Lawrence, Jr., 32, University of North Carolina 1941, the battalion S-3, had become the de facto commander.

Things were going even less well on the left flank and rear of the column. The Army provisional battalion, fragile to begin with, had fought itself out and was replaced by Harris' 3d Battalion, 7th Marines. By 2100 the Chinese had come down to within hand-grenade range of the trucks on the road. Harris deployed his George and Item Companies to push them back. Sometime before dawn, Lieutenant Colonel William Harris, son of Major General Field Harris, disappeared. He was last seen walking down the road with two rifles slung over his shoulder. A search for him found no body and it was presumed he had been taken prisoner. Major Warren Morris, the executive officer of the 1st Battalion, took over command of the 3d Battalion and it reached Koto-ri at about 0700 on Thursday morning.

Chinese prisoners taken along the road from Hagaru-ri to Koto-ri were identified as being from the *76th* and *77th Divisions* of *26th CCF Army*.

For Almond, most of Wednesday, 6 December, was absorbed with a visit by General J. Lawton Collins, the Army chief of staff. Collins and Almond dropped in at the command posts of the Army's 7th and 3d Infantry Divisions, but "weather precluded flying to Koto-ri" for a visit with Smith. Collins left at nightfall for Tokyo. The visit had gone well and Almond noted contentedly in his diary: "Gen. Collins seemed completely satisfied with the operation of X Corps and apparently was much relieved in finding the situation well in hand."

At Koto-ri, 7 December

First Lieutenant Leo R. Ryan, the adjutant of the 2d Battalion, 7th Marines, alarmed by Lieutenant Colonel Lockwood's apathy, pressed the battalion surgeon and assistant surgeon, Lieutenants (jg)

The role of the rifle companies in the breakout from Hagaru-ri on 6 December was to take the high ground on both sides of the road. Much fought over "East Hill" dominated the exit from Hagaru-ri. It was never completely taken, but the Chinese were pushed back far enough to permit the relatively safe passage of the division trains of vehicles.

Photo by Sgt Frank C. Kerr, Department of Defense Photo (USMC) A5465

Photo by Sgt Frank C. Kerr, Department of Defense Photo (USMC) A5466

Soldiers, readily recognizable as such in their short parkas, march in single file on 6 December along the MSR south of Hagaru-ri. RCT-31, badly mauled east of the reservoir and reduced in combat effectives to a small provisional battalion commanded by LtCol Berry K. Anderson, USA, was attached to the 7th Marines for the breakout.

Laverne F. Peiffer and Stanley I. Wolf, to examine him. Neither doctor was a psychiatrist, but they came to the conclusion that Lockwood was suffering from a neurosis that made him unfit for command. This was communicated to Colonel Litzenberg who confirmed Major Lawrence as the acting commander.

In mid-morning, Thursday, 7 December, to ease the passage of the division train, both the 2d and 3d Battalions, 7th Marines, were ordered to face about, move north again, and set up blocking positions on both sides of the road between Koto-ri and Hill 1182. On the way the 2d Battalion picked up 22 Royal Marine survivors who had been laagered up in a Korean house ever since Task Force Drysdale had passed that way. A VMO-6 pilot had spotted them three days earlier by the letters "H-E-L-P" stamped in the snow and had dropped rations and medical supplies.

Elsewhere on Thursday, 7 December, X Corps and Eighth Army had received orders from GHQ Tokyo to plan to withdraw, in successive positions if necessary, to the Pusan area. Eighth Army was to hold on to the Inchon-Seoul area as long as possible. X Corps was to withdraw through Hungnam and eventually to pass to the command of Eighth Army.

Almond visited Smith at Koto-ri and assured him that Soule's 3d Infantry Division would provide maximum protection from Chinhung-ni on into Hamhung. Smith

LtCol William F. Harris, commanding officer of the 3d Battalion, 7th Marines, and son of MajGen Field Harris, disappeared on the march from Hagaru-ri to Koto-ri. He was last seen moving down the road with two rifles slung over his shoulder. His exact fate remains a mystery.

National Archives Photo (USMC) 127-N-A45353

National Archives Photo (USA) 111-SC354254

General J. Lawton Collins, the Army chief of staff, left, visited X Corps on 6 December and warmly praised MajGen Almond, right, for his conduct of the battle. Collins got to the command posts of the 3d and 7th Infantry Divisions, but bad weather kept him from seeing MajGen Smith and the command post of the 1st Marine Division at Koto-ri.

was concerned over the coordination of artillery fire by the 3d Division. Almond promised that it would be under Marine control. He spoke briefly with Puller and Litzenberg and noted that night in his diary, "Morale is high in the Marine Division."

All elements of RCT-7 had closed into Koto-ri by 1700 on the evening of 7 December; but Division Train No. 1, which they were to have shepherded, did not get out of Hagaru-ri until 1600 on the 6th. A little more than a mile out of Hagaru-ri the Chinese came down on the column. They might have thought the train would be easy pickings; if so, they were wrong. They hit Major Francis "Fox" Parry's 3d Battalion, 11th Marines. The artillerymen, fighting as infantry, held them off. Another mile down the road and the process was repeated. This time the gunners got to use their howitzers, firing at pointblank range, and happily, if optimistically, guessed that they had killed or wounded all but about 50 of the estimated 500 to 800 attackers.

As the night wore on there was more fighting along the road. The division headquarters had a stiff scuffle sometime after midnight. The members of the division band were given the opportunity to demonstrate their skills as machine gunners. The Military Police Company was bringing out a bag of 160 able-bodied prisoners of war. The prisoners got caught between Chinese and American fires and most were killed. Night hecklers from David Wolfe's VMF(N)-513 helped and at dawn the omnipresent Corsairs from Frank Cole's VMF-312 came on station and resolved the situation. The column moved through the stark debris—there were still bodies lying about and many broken vehicles—of Hell Fire Valley and by 1000 on 7 December Division Train No. 1, after an all-night march, was in Koto-ri.

Eleven miles away in Hagaru, Division Train No. 2, unable to move onto the road until Division Train No. 1 had cleared, did not get started until well after dark on 6 December. At midnight, the head of the train was still barely out of the town. Lieutenant Colonel Milne, the train commander, asked for infantry help. Taplett's 3d Battalion, 5th Marines, was detailed to the job. Taplett moved forward with two companies. Nothing much happened until dawn on 7 December when the column was able to continue on under air cover.

In Hagaru-ri engineers and ordnance men were busy blowing up everything that could be blown up and burning the rest. Stevens' 1st Battalion, 5th Marines; Ridge's 3d Battalion, 1st Marines; and Drysdale's 41 Commando stood poised to leave but could not get out of town until Thursday morning, 7 December, after some fighting in Hagaru-ri itself, because of the clogged roads. Roise's 2d Battalion, 5th Marines, came off East Hill and fell in behind them as rear guard at about 1000. The Chinese once again seemed more interested in looting what was left of the town than in further fighting. After some light interference on the road, all elements of RCT-5 were safely tucked into the Koto-ri perimeter before midnight on the 7th.

A number of units—including Roise's 2d Battalion, 5th Marines; Ridge's 3d Battalion, 1st Marines; and Drake's 31st Tank Company—assert that they provided the rear point coming out of Hagaru-ri. Able Company engineers, however, busy with last-minute demolitions in the already burning town, probably have the best claim. In round figures, 10,000 Marines and soldiers, shepherding 1,000 vehicles, had marched 11 miles in 38 hours. Marine losses were 103 dead, 7 missing, and 506 wounded.

Marine engineers, arguably the greatest heroes of the campaign, had widened and improved the airstrip at Koto-ri so that it could handle World War II TBMs, no longer used as torpedo bombers, but now stripped-down utility aircraft that could bring in a few passengers—as many as nine—and lift out a corresponding number of wounded. The TBMs, plus the light aircraft and helicopters from VMO-6, took out about 200 casualties on 7 December and 225 more on the 8th. Most of the TBMs were piloted, not by squadron pilots, but by otherwise desk-bound aviators on the wing and group staffs.

National Archives Photo (USMC) 127-N-A5457

Chinese prisoners within the Hagaru-ri perimeter were herded into a stockade guarded by a detachment of Capt John H. Griffin's Military Police Company. When the Marines evacuated the town they left the wounded prisoners of war behind in a compound, telling them that their comrades would soon be down from the hills to help them.

March South from Koto-ri

There would be no rest at Koto-ri. By somebody's count 14,229 men had piled into Koto-ri, including the long-waited Army's 2d Battalion, 31st Infantry, which had arrived far too late to go forward to join its regiment, the shredded RCT-31, east of the reservoir. Reidy's battalion was to continue as part of Puller's RCT-1 in the break out.

Anderson's two-battalion collection of soldiers, quite separate from Reidy's battalion, had suffered additional casualties—both battle and from the cold—coming in from Hagaru-ri. Major Witte, one of the battalion commanders, was among the wounded. Anderson reorganized his shrinking command into two companies: a 31st Company under Captain George A. Rasula, a canny Finnish-American from Minnesota who knew what cold weather was all about, and a 32d Company under Captain Robert J. Kitz, who had been a company commander in Reilly's 3d Battalion, 31st Infantry, in Task Force Faith. Anderson then stepped aside from immediate command, giving the battalion to Major Robert E. Jones who had been Don Faith's S-1 and adjutant. As a paratrooper in World War II, Jones had jumped with the 101st Airborne Division near Eindhoven, Holland. Now, coming out of Koto-ri, his improvised battalion remained part of Litzenberg's RCT-7.

The Marines left Hagaru-ri in flames, wanting to leave no shelter for the Chinese. Veterans still argue as to which unit was the last to leave the town. Marines from Company A, 1st Engineer Battalion, charged with last minute demolitions, probably have the best claim to this honor.

National Archives Photo (USMC) 127-N-A5458

Photo by David Douglas Duncan

As the Marines moved south from Hagaru-ri to Koto-ri they had to pass through "Hell Fire Valley," site of Task Force Drysdale's heaviest losses. The Chinese had made no attempt to salvage the abandoned vehicles, but the Marines did and learned that some needed nothing more than a push to start them.

The march south was to be resumed at first light on Friday, 8 December. It would be a "skin-the-cat" maneuver with the rifle companies leap-frogging along the high ground on each side of the road while the heavily laden vehicles of the division trains made their way toward Funchilin Pass and then down the pass to Chinhung-ni. At the foot of the pass the Marines could expect to find elements of the Army's 3d Infantry Division manning the outer defenses of the Hamhung-Hungnam area. But the road was not yet open. Smith had been warned, as early as 4 December, that the Chinese had blown a critical bridge halfway down the pass. Here water came out from the Changjin Reservoir through a tunnel into four giant pipes called "penstocks." The bridge had crossed over the penstocks at a point where the road clung to an almost sheer cliff. If the division was to get out its tanks, artillery, and vehicles the 24-foot gap would somehow have to be bridged.

Lieutenant Colonel Partridge, the division engineer, had made an aerial reconnaissance on 6 December and determined that the gap could be spanned by four sections of an M-2 steel "Treadway" bridge. He had no such bridge sections, but fortuitously there was a detachment of the Treadway Bridge Company from the Army's 58th Engineer Battalion at Koto-ri with two Brockway trucks that could carry the bridge sections if they could be air-delivered. A section was test-dropped at Yonpo by an Air Force C-119 and got smashed up in the process. Not discouraged, Partridge pressed for an airdrop of eight sections—to give himself a 100 percent insurance factor that at least four sections would land in useable condition. The 2,500-pound bridge sections began their parachute drop at 0930 on 7 December. One fell into the hands of the Chinese. Another was banged up beyond use. But six sections landed intact. Plywood center sections for wheeled traffic were also dropped. Next, the Brockway trucks would have to deliver the sections to the bridge site three-and-a-half miles away, a location likely to be defended fiercely by the Chinese.

Partridge met with Litzenberg and it was decided that the Brockway trucks would move at the front of the 7th Marines' regimental train after RCT-7 jumped-off at 0800 on 8 December. The bridge site was dominated by Hill 1081 so Lieutenant Colonel Schmuck's 1st Battalion, 1st Marines, at Chinhung-ni was

Photo by Sgt William R. Keating, Department of Defense Photo (USMC) A5461

One of the vehicles salvaged by the Marines in "Hell Fire Valley" was a bullet-ridden Army ambulance. Battered, but still in running condition, the Marines used it to evacuate casualties incurred along the line of march. Once safely in Masan, a large number of Army vehicles were returned to the Army.

ordered to advance overland three miles to the north to take the hill. All of this required exquisite timing.

First objective for Litzenberg's 7th Marines coming out of Koto-ri was the high ground on the right of the road for a distance of about a mile-and-a-half. Murray's 5th Marines would then pass through the 7th Marines and take and hold the high ground for the next mile. Puller's 1st Marines was to stay in Koto-ri until the division and regimental trains had cleared and then was to relieve the 5th and 7th Marines on their high ground positions so the trains could pass on to Funchilin Pass. The 5th and 7th Marines, relieved by the 1st, would then move on down the pass toward Hamhung. The 11th Marines artillery would displace from battery firing position to battery firing position but for much of the time would be limbered up and on the road. Heavy reliance for fire support would be placed on the Corsairs and organic mortars. Tanks would follow at the end of the vehicular column so there would be no chance of a crippled tank blocking the road.

Task Force Dog, under Brigadier General Mead and consisting of the 3d Battalion, 7th Infantry, liberally reinforced with tanks and artillery, had started north on 7 December, passed through Su-dong, and by late afternoon had reached Chinhung-ni. Schmuck's 1st Battalion, 1st Marines, after being relieved by Task Force Dog, moved into an assembly area several miles north of Chinhung-ni.

The jump-offs from both Koto-ri and Chinhung-ni on the morning of 8 December were made in a swirling snowstorm. Schmuck's Marines started the six-mile march up the MSR to the line of departure at 0200. His plan was for Captain Robert P. Wray's Company C to take Hill 891, the southwestern nose of Hill 1081, and hold it while his other two rifle companies passed through and continued the attack. Captain Barrow's Company A was to attack east of the road and on up to the summit of Hill 1081. Captain Noren's

Koto-ri as it looked on 8 December, the day that the march to the south continued. Virtually all the combat strength of the 1st Marine Division, plus some Army troops, had concentrated there for the breakout. The next critical terrain feature would be Funchilin Pass where a blown bridge threatened to halt the march. Its repair posed a problem for Marine engineers.

Photo by Sgt William R. Keating, Department of Defense (USMC) A5354

Photo by Sgt William R. Keating, Department of Defense Photo (USMC) A5361

Snow-dusted M-4A3 Sherman tanks of LtCol Harry T. Milne's 1st Tank Battalion await the word at Koto-ri to move out to the south. The tanks had turned in a disappointing performance with Task Force Drysdale. There would be further problems with the tanks in Funchilin Pass.

Immediately outside of Koto-ri, two Chinese soldiers willingly surrender to members of a Marine rifle company early on 9 December. Each leg of the withdrawal, from Yudam-ni to Hagaru-ri, from Hagaru-ri to Koto-ri, and from Koto-ri to Chinhung-ni, showed a marked improvement in Marine tactics to deal with the situation.

Photo by Cpl Peter W. McDonald, National Archives Photo (USMC) 127-N-A5388

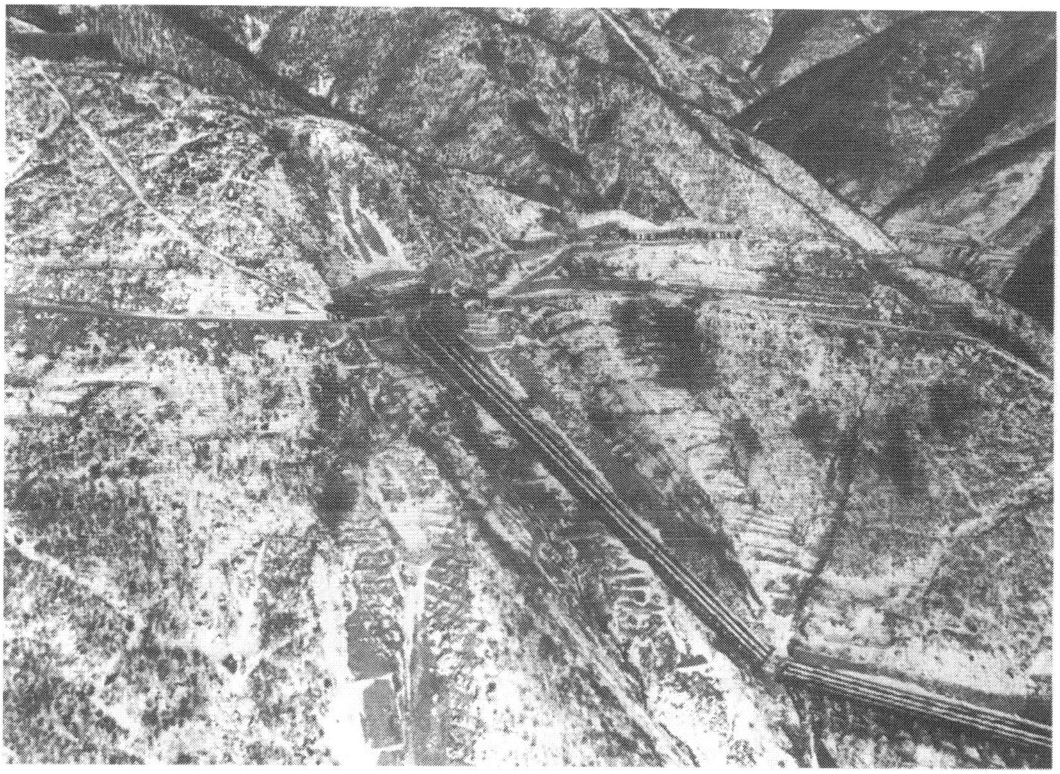

Photo by SSgt Ed Barnum, Department of Defense Photo (USMC) A130504

Aerial view taken from one of VMO-6's light observation aircraft, flown by 1stLt John D. Cotton, shows the power station, the pipes or "penstocks" that carried off the water, and the precarious nature of the road occupying a thin shelf cut into the precipitous slope.

Company B would be on the left flank, moving along the slope between Barrow and the road.

Wray had his objective by dawn. On it Schmuck built up a base of fire with his 81mm mortars and an attached platoon of 4.2-inch mortars—the effective, but road-bound "four-deuces." Also effective, but tied to the road, were five Army self-propelled anti-aircraft guns—quad .50-calibers and duel 40mms—attached from Company B, 50th Antiaircraft Artillery (Automatic Weapons) Battalion.

Things went like clockwork. Schmuck's main attack jumped off at 1000. Barrow clambered up the hogback ridge that led to the summit of Hill 1081; Noren advanced

A mixed group of Marines and soldiers struggle up an ice-covered slope somewhere south of Koto-ri. The weather and the terrain were at least as much of an enemy as the Chinese. Marines, disdainful of the Army's performance east of the reservoir, learned in the march-out from Hagaru-ri that soldiers, properly led, were not much different from themselves.

Photo by Cpl Peter W. McDonald, National Archives Photo (USMC) 127-N-A5389

Sketch by Sgt Ralph Schofield, USMCR

Going downhill was easier for the most part than going up, but wherever it was the march was single file of Marines, or at best a double file. Even stripped down to essentials, the average Marine carried 35 to 40 pounds of weapons, ammunition, rations, and sleeping bag. Anything more, such as toilet articles, shelter half, or poncho, was a luxury.

along the wooded western slope of the hill. Noren met scattered resistance, was stopped momentarily by two enemy machine guns, which he then took out with a tidy schoolbook solution—engaging the enemy with his own machine guns and 60mm mortars while a platoon worked around in a right hook. He then ran into a bunker complex, took it after a savage fight, and found a kettle of rice cooking in the largest bunker. Schmuck moved his headquarters forward and set up his command post in the bunker only to find it louse-ridden. The day cost Noren three killed and six wounded.

Barrow had gone up the ridge against no enemy whatsoever, impeded only by the icy ridgeline, so narrow that he had to march in a dangerous single file. Through a break in the snowstorm, Barrow got a glimpse of a strongly bunkered Chinese position on a knob between his company and the crest of the hill. He elected to do a double envelopment, sending his 2d Platoon around to the left and his 1st Platoon around to the right. He went himself with the 3d Platoon up the center in a frontal attack. It all came together in a smashing assault. Barrow's

Fresh snow fell during the march from Koto-ri to the top of Funchilin Pass. When the column on the road halted, as it frequently did, the Marines tended to bunch up, making themselves inviting targets for Chinese mortar and machine gun fire. March discipline had to be enforced by tough corporals and sergeants more than by orders from the top.

Photo by Sgt Frank C. Kerr, National Archives Photo (USMC) 127-N-A5358

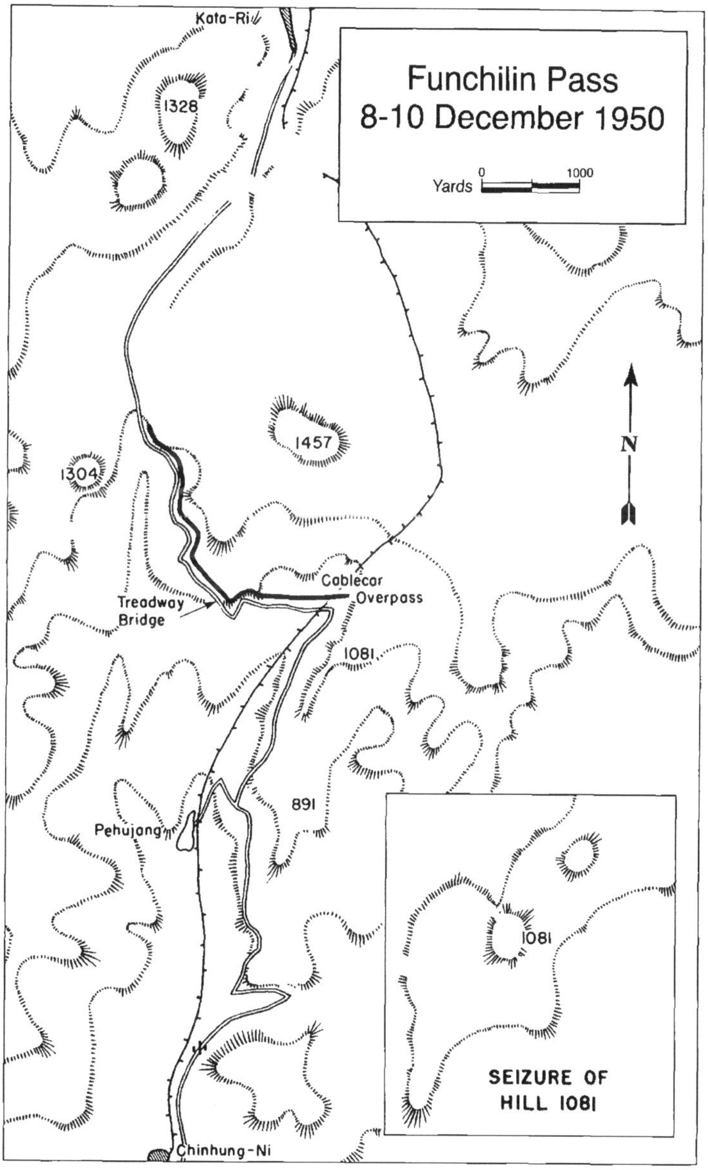

Marines counted more than 60 Chinese dead. They themselves lost 10 killed, 11 wounded. The snow ended and the night was clear. At midnight a Chinese platoon bravely but foolishly tried to evict Barrow's Marines and lost 18 killed. To Barrow's left, all was quiet in front of Noren's position.

To their north, Litzenberg's 7th Marines had come out of Koto-ri on schedule on the morning of 8 December. Counting the Army provisional battalion he had four battalions. Two were to clear each side of the road. One was to advance along the MSR, to be followed by the regimental train and the reserve battalion. Major Morris had been assigned to take Hill 1328 on the right of the road with his 3d Battalion. Going was slow. By mid-morning Litzenberg grew impatient and urged him to commit his reserve company. Morris snapped back: "All three companies are up there—50 men from George Company, 50 men from How, 30 men from Item. That's it." Shortly after noon, Litzenberg committed his regimental reserve, Lawrence's 2d Battalion, to come to the assistance of Morris. By nightfall the two battalions had joined but not much more was accomplished.

Left of the road, the provisional Army battalion, under Major Jones, had jumped off on time and, with the help of two Marine tanks, had moved along against light resistance. In two jumps Jones reached Hill 1457 where his soldiers dug in for the night. Their position was raked by Chinese automatic fire, and in a brief nasty action 12 enemy were killed at a cost of one soldier killed, four wounded.

Litzenberg's executive officer, Lieutenant Colonel Frederick Dowsett, had been shot through the ankle the day before. Litzenberg moved Raymond Davis up to executive officer to replace him and gave the 1st Battalion, still his strongest battalion, to Major Sawyer, whose wound had proved superficial. Sawyer's initial mission was to move a mile down the road and wait for the 3d Battalion to come up on his right flank. The 1st Battalion now had it own fight.

Sawyer's lead platoon came under fire from Hill 1304. Baker Company continued to move against the high ground just left of the road while Able and Charlie Companies moved more deeply to the right against the hill. Baker Company was caught in a crossfire; the company commander, Lieutenant Kurcaba, was killed, two of his platoon leaders were

Photo by Sgt Frank C. Kerr, National Archives Photo (USMC) 127-N-A5357

A section of two 81mm mortars set up in the snow to give fire support for a rifle company in the attack. These high-angle medium mortars, with a shell almost as lethal as a 105mm howitzer round, were considered to be the infantry battalion commander's own "artillery." Mortars were particularly effective in the defense of a perimeter.

wounded. First Lieutenant William W. "Woody" Taylor took over command of the company and had his objective by nightfall. Able and Charlie Companies meanwhile had taken Hill 1304 without much trouble. Sawyer divided his battalion into two perimeters for the night. Vehicular movement along the MSR was halted.

It had been nearly noon on 8 December before Murray's 5th Marines, following behind the 7th Marines, moved out of Koto-ri. Stevens' 1st Battalion was in the lead. Stevens sent out his Baker and Charlie Companies to take Hill 1457. Charlie Company joined up unexpectedly with the Army's provisional battalion and the soldiers and Marines had the Chinese off the high ground by mid-afternoon. Baker and Charlie Companies, combined with the Army troops, formed a perimeter for the night. Able Company had its own perimeter closer to the MSR. Murray moved 41 Commando, in reserve, up behind the 1st Battalion.

Meanwhile, the 2d and 3d Battalions of Puller's 1st Marines held Koto-ri itself. For the defenders the problem was not the scattered small arms fire of the Chinese, but the flood of civilian refugees coming down the road from the north. They could not be admitted into the perimeter because of the probability that the Chinese had infiltrated them. During the bitterly cold night, two babies were born with the help of Navy doctors and corpsmen. In all their misery these thousands of civilians had to wait outside the lines until Koto-ri was vacated. They then followed behind the Marines, as best they could, until the presumed safety of Hamhung-Hungnam might be reached.

During the day Smith, always conscious of his dead, attended a funeral at Koto-ri. What had been an artillery command post, scraped more deeply into the frozen ground by a bulldozer, became a mass grave. A total of 117 bodies, mostly Marines but some soldiers and Royal Marines, were lowered into the hole. A Protestant and a Catholic chaplain officiated. The bulldozer covered the bodies with a mound of dirt.

Sergeant Robert Gault, head of the Graves Registration Section of the 7th Marines, remembered the funeral this way:

> We had a chaplain of each faith, and the fellows had made a big hole and laid the fellows out in rows the best

Task Force Dog included these self-propelled 155mm howitzers shown in firing position covering Funchilin Pass on 10 December near Chinhung-ni. Marine 155mm howitzers at this time were still tractor-drawn and some were lost coming out of Yudam-ni. At a greater distance an Army tank can be glimpsed.

Photo by TSgt James W. Helms, Jr., Department of Defense Photo (USMC) A156320

Photo by Cpl James Lyle, National Archives Photo (USA) 111-SC354456

A squad-sized Army patrol, led by Sgt Grant J. Miller, from the 3d Infantry Division's Task Force Dog, moves up from Chinhung-ni into Funchilin Pass on 9 December. The tank at the side of the road appears to be a Soviet-built T-34 knocked out a month earlier by the 7th Marines.

> we could and put ponchos over them. As soon as each chaplain had said his little bit for the fellows, we would cover them up and close them in. Everyone was given—I think under the circumstances—a very fine burial. It wasn't like the one back at Inchon and Hungnam. It wasn't like the one where we had crosses for the boys painted white and all the preliminaries: flowers that we could get for them—we'd go out and pick them. It wasn't like that, no. It was one where we were just out in a field, but it was one with more true heart.

There was more snow during the night, but Saturday, 9 December, dawned bright, clear, and cold. South of Funchilin Pass, Noren moved his Baker Company to the next high ground to his front and Barrow had his Able Company test-fire their weapons before beginning the assault of Hill 1081. Barrow then attacked in a column of platoons behind a thunderous preparation by close air, artillery, and mortars. Even so his lead platoon, under First Lieutenant William A. McClelland, was hard hit as it moved forward by rushes, stopping about 200 yards from the crest. Under cover of air strikes by four Corsairs and his own 60mm mortars, Barrow moved his 2d and 3d Platoons forward and by mid-afternoon his Marines had the hill. The two-day battle cost Barrow almost exactly half his company. He had started up the hill with 223 Marines; he was now down to 111 effectives. But 530 enemy dead were counted and the Marines held the high ground commanding Funchilin Pass.

On the MSR that Saturday, moving south from Koto-ri, the 7th Marines resumed its attack. The rest of Hill 1304 was taken. Captain John Morris with his Company C and a platoon from Company B moved down the road and secured the bridge site. The rest of Company B, following behind, overran an enemy position garrisoned by 50 Chinese so frozen by the cold that they surrendered without resistance.

The old war horse, General Shepherd, arrived from Hawaii the day before on what was his fifth trip to Korea, this time as "Representative of Commander Naval Force, Far East, on matters relating to the Marine Corps and for consultation and advice in connection with the contemplated amphibious operations now being planned." Shepherd may have thought he had more authority than he really had. In his 1967 oral history he said:

> When reports came back that the cold weather had set in and they weren't able to make the Yalu River and things began falling apart, Admiral Radford sent me to Korea—I think [the orders] came from the Chief of Naval Operations on the recommendation of Admiral Joy—that [I] was to take charge of the evacuation of the Marines from Hungnam.

More accurately, the Chief of Naval Operations, Admiral Forrest P. Sherman, had probably been prompted by back-channel messages to Admiral Radford to send Shepherd to the Far East "for the purpose of advising and assisting Commander Naval Forces, Far East [Admiral Joy], with particular emphasis on Marine Corps matters."

Shepherd did recognize that there could be a conflict of command because of Almond's actual command of X Corps. On arriving in Tokyo on 6 December he met with General MacArthur and noted "General MacArthur was unqualified in his admiration and praise for the effective contribution which Marines had made throughout the whole of the Korea fighting. His general demeanor [however] was not one of optimism."

After more conferences and meetings in Tokyo, Shepherd left on 8 December for Hungnam and on

Photo by Sgt Frank C. Kerr, Department of Defense Photo (USMC) A5370

Litzenberg's 7th Marines led the way out of Koto-ri at dawn on 8 December. Murray's 5th Marines followed the 7th Marines and in turn was followed by two battalions of Puller's 1st Marines. Infantry units, moving from hilltop to hilltop on both sides of the road covered the movement of the division trains. Some called the maneuver "hop-scotch," others called it "skinning-the-cat."

arrival went immediately to the *Mount McKinley*. Here on the next day he attended a meeting on outloading and naval support also attended by Admirals Joy, Struble, and Doyle, and General Harris. A press conference followed. Shepherd praised the operations of the X Corps and said that he was there to assist General Almond. He was anxious to get up to the reservoir to see things for himself. He made the trip in a TBM, landing at Koto-ri after circling Hagaru-ri. He then met with Smith for an hour or more. Smith told him that all casualties would be out by the end of the day. Shepherd announced that he intended to march out with the division. Smith dug in his heels and said absolutely not.

Shepherd returned to the airstrip. A number of war correspondents, among them "Maggie" Higgins, Keyes Beach, and the photographer David Douglas Duncan, had wangled their way to Koto-ri. While Shepherd's plane was warming up, Colonel Puller arrived leading Higgins by the hand. Puller said: "General Smith says take this woman out of his hair and see that she goes out on your plane." Shepherd turned to Higgins, whom he had met at Inchon, saying, "Maggie, it's too bad. I wanted to march down too." The plane completed its loading of wounded and taxied to the end of the strip. It was dusk and as the plane took off Shepherd could see machine gun tracer bullets reaching up at the underside of

As the march went on, cold-benumbed Chinese soldiers surrendered in increasing numbers. This group, probably the remnants of a platoon or perhaps a company, surrendered to Company C, 1st Battalion, 7th Marines, south of Koto-ri on 9 December.

Photo by Sgt Frank C. Kerr, National Archives Photo (USMC) 127-N-A5377

Two Chinese, anxious to surrender, get a quick pat-down search for weapons by members of Company C, 7th Marines, but there was no fight left in them. Once given a cigarette and perhaps a chocolate bar by their captors, they would follow along uncomplainingly into eventual captivity.

National Archives Photo (USMC) 127-N-A5378

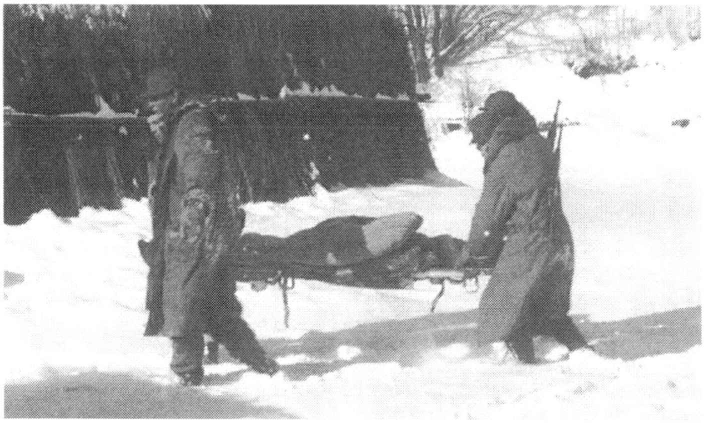

Photo by Cpl Peter W. McDonald, National Archives Photo (USMC) 127-N-A5379

First step in evacuation of casualties, alive or dead, during the march-out was to get them down to the road by stretcher. Here four Marines, without helmets or packs, carry a stretcher down from the high ground somewhere south of Koto-ri on 9 December. Fresh snow had fallen the previous night.

the plane. Leaning over to Higgins, the irrepressible Shepherd said to her, "If we get hit, we will die in each other's arms."

From Koto-ri to Chinhung-ni

The column of division vehicles, protected on both sides by the Marine infantry in the hills, crawled along the road south of Koto-ri at a snail's pace and with frequent stops. The Marines, who watched the crawling column from their perches in the hills, wondered profanely why the vehicles had to be piled high with tent frames, wooden doors, and other luxuries of life.

Partridge had held back the Army's Brockway trucks, with their precious cargo of bridge sections, in Koto-ri until first light on 9 December when he considered the MSR secure enough for him to move them forward. He then joined Sawyer's 1st Battalion at the head of the column. Everything worked at the bridge site like a practiced jigsaw puzzle. Army and Marine engineers rebuilt the abutments with sandbags and timbers. A Brockway truck laid the steel treadways and plywood deck panels. At noon, Almond flew overhead in his "Blue Goose" to see for himself that things were going well. Installation was done in three hours and at 1530 Partridge drove his jeep back to the top of the pass to tell Lieutenant Colonel Banks that he could bring Division Train No. 1 down the defile. The first vehicles began to cross the bridge at about 1800. Sawyer's Marines kept the enemy at a distance and captured 60 prisoners in the process. All night long vehicles passed over the bridge.

At 0245 on Sunday morning, 10 December, the head of the column reached Chinhung-ni. Colonel Snedeker, the division's deputy chief of staff, had positioned himself there to direct the further movement of the vehicle serials. The 7th Marines followed Division Train No. 1 down the pass. Up on the plateau Ridge's 3d Battalion, 1st Marines, had come out of Koto-ri and relieved the 3d Battalion, 7th Marines, on Hill 1328 where it had a fight with

Frozen corpses are unloaded from a truck at Koto-ri where they will be buried in a mass grave. A 155mm howitzer can be seen in the background. The dead, 117 of them, mostly U.S. Marines but some soldiers and Royal Marines, were interred in a hole originally bull-dozed into the ground to serve as an artillery fire direction center.

Photo by Sgt Frank C. Kerr, National Archives Photo (USMC) 127-N-A5366

National Archives Photo (USMC) 127-N-A402841

LtGen Shepherd, right, Commanding General, Fleet Marine Force, Pacific, arrived on 8 December on his fifth trip to Korea. Although not in the operational chain-of-command, he had arrived to oversee the evacuation of the Marines. RAdm Doyle, left, and MajGen Harris, center, greet Shepherd. All three look glum.

about 350 resurgent Chinese. At 1030, General Smith closed his command post at Koto-ri and flew to his rear command post at Hungnam.

Puller brought out the remainder of RCT-1 from Koto-ri on the afternoon of the 10th. Milne's tanks, including the tank company from the 31st Infantry, followed behind the elements of RCT-1 on the road. Ridge's 3d Battalion was already deployed on the high ground on both sides of the MSR south of Koto-ri. The plan was for Sutter's 2d Battalion to relieve Stevens' 1st Battalion, 5th Marines, on Hill 1457.

As the last Americans left Koto-ri, the Army's 92d Field Artillery, firing from Chinhung-ni, shelled the town with its long-range 155mm guns. There was confusion at the tail of the column as Korean refugees pressed close. The tankers fired warning shots to make them stay back. Panic developed as the rumor spread that the Marines were shooting the refugees. The tanks passed on down the road, protected on both sides at first by Ridge's Marines in the hills. But Sutter, having begun his climb up Hill 1457 and finding it a long way off and with no enemy in sight, asked Puller's permission, which he received, to return to the road.

Ridge pulled his companies off Hill 1304 and the high ground on the opposite side of the MSR at about 2100. Ridge's battalion was the last major unit to descend the pass, following behind Jones' provisional battalion of soldiers and the detachment of the 185th Engineers. Harry Milne's tanks were behind Ridge with no infantry protection except the lightweight division Reconnaissance Company mounted in jeeps. It was now about midnight.

By then both division trains, all of RCT-7, and most of the 11th Marines had reached Chinhung-ni. The 5th Marines followed the 7th Marines. Beyond Chinhung-ni, guerrillas were reported to be active in the vicinity of Sudong, but the division trains and both the 5th and 7th Marines passed through without interference. Some time after midnight when the vehicles of RCT-1 reached the town sudden swarms of Chinese came out of the houses of the village with burp guns and grenades. Truck drivers and casuals, both Army and Marine, fought a wild, shapeless action. Lieutenant Colonel John U. D. Page, an Army artillery officer, took charge, was killed, and received a posthumous Medal of Honor. Lieutenant Colonel Waldon C. Winston, an Army motor transport officer, took his place. It was dawn before the place was cleaned up. RCT-1 lost nine trucks and a personnel carrier; 8 men killed and 21 wounded.

Meanwhile, Milne's tanks, some 40 of them, descending the narrow, icy-slick road of Funchilin Pass had run into trouble. About a mile short of the Treadway bridge the brakes of the ninth tank from the end of the column froze up. The tanks to its front clanked on, but the immobile ninth tank blocked the eight tanks to the rear. Close behind came the refugees. Left guarding the nine tanks was First Lieutenant Ernest C. Hargett's 28-man reconnaissance platoon. Five Chinese soldiers emerged from the mass of refugees and one, in English, called upon Hargett to surrender. Hargett, covered by a BAR-man, approached the five Chinese cautiously. The English-speaking one stepped aside and the four others produced burp guns and grenades. A grenade wounded Hargett. His BAR-man, Corporal George A. J. Amyotte, cut the five Chinese down, but more Chinese materialized on the road and the steep slope of the hill. Hargett backed away with his platoon. The last tank in the column was lost to the

Department of Defense Photo (USMC) A5376

The Chinese had blown a critical bridge halfway down Funchilin Pass where water flowed downward from Changjin Reservoir and passed through four giant pipes called "penstocks." As early as 4 December, MajGen Smith knew that this 24-foot gap would have to be bridged if his vehicles were to reach Hungnam.

Chinese. Meanwhile, the crew of the tank that blocked the road had succeeded in freeing the frozen brakes and was ready to proceed. But the crews of the remaining seven tanks had departed, leaving the hatches of their tanks open. A member of Hargett's platoon, who had never driven a tank, managed to bring out one tank. The night's adventure cost Hargett two men killed and 12 wounded.

Engineers were waiting at the Treadway bridge, ready to blow it up. They thought the two tanks and Hargett's platoon were the last to come by. They blew the bridge, but one Marine had been left behind. Private First Class Robert D. DeMott from Hargett's platoon had been blown off the road by a Chinese explosive charge. Regaining consciousness, he got back on the road and joined the refugees.

Marines from Litzenberg's regiment, along with some attached soldiers, on 9 December reached the blown bridge in Funchilin Pass, which, unless replaced, would stop any further southward movement of wheeled or tracked vehicles. Plans were already afoot to bridge the gap with a Treadway bridge to be airdropped in sections at Koto-ri.

Photo by Sgt Frank C. Kerr, Department of Defense Photo (USMC) A5375

Photo by Sgt William R. Keating, Department of Defense Photo (USMC) A5408

By evening on 9 December the Treadway bridge was in place and men and vehicles could move unimpeded down the MSR through Funchilin Pass. From here on enemy resistance was limited to small-scale firefights and ambushes. The most sizable resistance would come near Sudong.

The wind that blew from Manchuria and beyond, "down over the Yalu and the mountains all around. . . down into the gorges with their frozen streams amd naked rocks. . . down along the ice-capped road—now shrieking and wild—that wind," said noted photographer David Douglas Duncan, "was like nothing ever known by the trapped Marines, yet they had to march through it."

Photo by David Douglas Duncan

Photo by Sgt Frank C. Kerr, Department of Defense Photo (USMC) A5372

Marines march along a particularly precipitous portion of the road winding down through Funchilin Pass on 9 December. The day before had seen the launching of an exquisitely timed maneuver—the exit of the 5th and 7th Marines from Koto-ri and the simultaneous advance of the 1st Battalion, 1st Marines, from Chinhung-ni to take the high ground controlling the pass.

He heard the detonation that blew the bridge, but figured that he could make his way on foot through the gatehouse above the penstocks. This he did as did many of the following refugees.

Warm Welcome at Hungnam

Donald Schmuck, from his position on Hill 1081, watched the lights of the tanks descending the pass and at 0300 gave orders for Barrow's Company A to begin its withdrawal. At 1300 on 11 December the last units of the division passed through Chinhung-ni. By 1730 they had gone through Majon-dong and by 2100 most had reached their Hamhung-Hungnam assembly areas. They found a tent camp waiting for them. Lieutenant Colonel Erwin F. Wann, Jr.'s 1st Amphibian Tractor

Coming down Funchilin Pass on 10 December, the Marine column was intermixed with many Korean refugees fleeing the Chinese. Numbers were such as to interfere with military traffic. Behind them, Yudam-ni, Hagaru-ri, and Koto-ri were left as deserted ghost towns.

Photo by Sgt William R. Keating, National Archives Photo (USMC) 127-N-A5407

National Archives Photo (USMC) 127-N-A5901

LtCol John H. Partridge, whose engineers performed miracles, particularly in scraping out the airstrip at Hagaru-ri and installing the airdropped Treadway bridge in Funchilin Pass, received a second Bronze Star from MajGen Smith at Masan in early January 1951.

Battalion had done much of the preparation for their arrival. Chow lines were open for the continuous serving of hot meals. Wann, 31 and Naval Academy 1940, had been an amphibian tractor officer at Bougainville, Guam, and Iwo Jima. The weather seemed almost balmy after the unrelieved subzero temperatures of the plateau. Milne's tanks continued on to the LST staging area, arriving just before midnight. From 8 through 11 December, the division had lost 75 men dead, 16 missing, and 256 wounded, for a total of 347 casualties.

As late as Saturday, 9 December, General Smith believed that the 1st Marine Division, once concentrated, would be given a defensive sector to the south and southwest of Hungnam. A day earlier his deputy chief of staff, Colonel Snedeker, who was running his rear headquarters, issued tentative orders for Puller's RCT-1 to organize defensively at Chigyong, with Murray's RCT-5 and Litzenberg's RCT-7 preparing to defend Yonpo airfield.

But on that Saturday, Almond received his formal orders from MacArthur to redeploy X Corps to South Korea and Smith learned that his division would be loading out immediately on arrival. At this point Almond regarded the 1st Marine Division as only marginally combat effective. He considered the 7th Infantry Division, except for its loss of almost a complete regimental combat team, to be in better condition. In best condition, in his opinion, was the 3d Infantry Division, which he visited almost daily.

Almond therefore decided that once the 1st Marine Division passed through the Hamhung-Hungnam perimeter defense it would be relieved from active combat and evacuated. Second priority for evacuation would be given the 7th Infantry Division. Last out would be the 3d Infantry Division.

The Hungnam-Hamhung defensive perimeter, as neatly drawn on the situation maps in Almond's headquarters, consisted of a main line of resistance (MLR) about 20 miles long arcing in a semicircle from north of Hungnam around to include Yonpo. In front of the MLR was a lightly held outpost line of resistance. The northernmost sector, beginning at the coastline, was given to Major General Kim Pak Il's ROK I Corps, which, having arrived uneventfully from Songjin, began moving into line on 8 December. The lift-off from Songjin by LSTs, merchant ships, and the attack transport USS *Noble* (APA 218) had been completed in three days. Counterclockwise, next in line on the perimeter, came Barr's 7th Infantry Division with two sound regiments, followed by Soule's 3d Infantry Division. The southern anchor of the perimeter was held by the 1st Korean Marine Corps Regiment, which had the mission of defend-

LtCol Harry T. Milne, commanding officer of the 1st Tank Battalion, in a post-war photograph. Tanks were vital to the Marine breakout, but their record was marred by poor performance with Task Force Drysdale and, later, in Funchilin Pass.

National Archives Photo (USMC) 127-N-A552011

Photo by Sgt William R. Keating, National Archives Photo (USMC) 127-N-A5404

Once the high ground commanding Funchilin Pass had been taken and the Treadway bridge was in place, the 1st Marine Division could descend almost unimpeded to the sea. All day long on 10 December Marine troops marched down the pass until they reached first Chinhung-ni and then went beyond Sudong-ni where they took trucks to Hamhung.

ing the Yonpo airfield. As the evacuation progressed, the MLR was to shrink back to successive phase lines.

Admiral Doyle, as commander, Task Force 90, assumed control of all naval functions on 10 December. Marine Colonel Edward Forney, a X Corps deputy chief of staff whose principal duties were to advise Almond on the use of Marine and Navy forces, was now designated as the Corps' evacuation control officer. The Army's 2d Engineer Special Brigade would be responsible for operating the dock facilities and traffic control. A group of experienced Japanese dock workers arrived to supplement their efforts,

Years later Almond characterized Forney's performance as follows: "I would say that the success of [the evacuation] was due 98 percent to common sense and judgment and that this common sense and judgment being practiced by all concerned was turned over to General Forney who organized the activities in fine form. I mean *Colonel* Forney, he should have been a *General*!"

General Field Harris briefed General Shepherd on 10 December on the status of the 1st Marine Aircraft Wing. Ashore were Marine fighter squadrons -312, -542, -513, and -311. Afloat were VMF-212 in the *Bataan*, VMF-214 in the *Sicily*, and VMF-323 in the *Badoeng Strait*. Shepherd learned that the wing had been offered either the K-10 airfield near Masan or K-1 near Pusan. K-1 was preferable because it was the better field and close to Pusan's port facilities.

The 11th of December was a busy day. X Corps issued its Operation Order 10-50, calling for the immediate embarkation of the 1st Marine Division. The perimeter would shrink progressively as then the 7th and 3d Infantry Divisions, in turn, were withdrawn. As the perimeter contracted, naval gunfire and air support would increase to defend the remaining beachhead. General MacArthur himself arrived that day at Yonpo, met with Almond, and approved the X Corps evacuation plan. He told Almond that he could return to

(Continued on page 118)

A Marine, left, and a Korean soldier, right, check the meager baggage of a Korean family on 10 December at some point east of Chinhung-ni before allowing the family to proceed to Hungnam. Some 91,000 refugees, driven by hunger, the cold, and fear of the Chinese, would be evacuated.

Photo by Cpl James Lyles, National Archives Photo (USA) 111-SC354459

Medals of Honor

The Medal of Honor, the Nation's highest award for valor, has been given to 294 Marines since its inception in 1862. The Korean War saw 42 Marines so honored. Of this number, 14 awards were made for actions incident to the Chosin Reservoir campaign. Seven of these awards were posthumous.

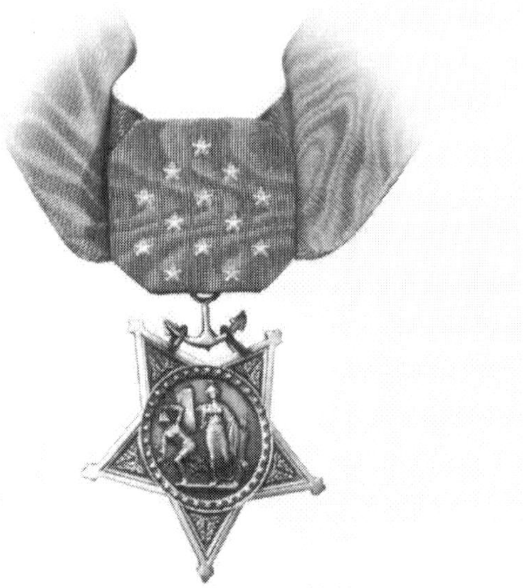

Staff Sergeant Archie Van Winkle

Staff Sergeant Archie Van Winkle, 25, of Juneau, Alaska, and Darrington, Washington, a platoon sergeant in Company B, 1st Battalion, 7th Marines, awarded the Medal of Honor for gallantry and intrepidity in action on 2 November 1950 near Sudong wherein he led a successful attack by his platoon in spite of a bullet that shattered his arm and a grenade that exploded against his chest.

Sergeant James I. Poynter

Sergeant James I. Poynter, 33, of Downey, California, a squad leader with Company A, 1st Battalion, 7th Marines, posthumously awarded the Medal of Honor for his actions on 4 November south of Sudong, where, although already critically wounded, he assaulted three enemy machine gun positions with hand grenades, killing the crews of two and putting the third out of action before falling mortally wounded.

Corporal Lee H. Phillips

Corporal Lee H. Phillips, 20, of Ben Hill, Georgia, a squad leader with Company E, 2d Battalion, 7th Marines, posthumously awarded the Medal of Honor for actions on 4 November 1950 near Sudong where he led his squad in a costly but successful bayonet charge against a numerically superior enemy. Corporal Phillips was subsequently killed in action on 27 November 1950 at Yudam-ni.

Second Lieutenant Robert D. Reem

Second Lieutenant Robert D. Reem, 26, of Elizabethtown, Pennsylvania, a platoon leader in Company H, 3d Battalion, 7th Marines, posthumously awarded the Medal of Honor for actions on 6 November 1950 near Chinhung-ni. Leading his platoon in the assault of a heavily fortified Chinese position, he threw himself upon an enemy grenade, sacrificing his life to save his men.

First Lieutenant Frank N. Mitchell

First Lieutenant Frank N. Mitchell, 29, of Indian Gap, Texas, a member of Company A, 1st Battalion, 7th Marines, posthumously awarded the Medal of Honor for extraordinary heroism in waging a single-handed battle against the enemy on 26 November 1950 near Yudam-ni to cover the withdrawal of wounded Marines, notwithstanding multiple wounds to himself.

Staff Sergeant Robert S. Kennemore

Staff Sergeant Robert S. Kennemore, 30, of Greenville, South Carolina, a machine gun section leader with Company E, 2d Battalion, 7th Marines, awarded the Medal of Honor for extraordinary heroism during the night of 27-28 November 1950 north of Yudam-ni in deliberately covering an enemy grenade whose explosion cost him both of his legs.

Private Hector A. Cafferata, Jr.

Private Hector A. Cafferata, Jr., 21, born in New York City, a rifleman with Company F, 2d Battalion, 7th Marines, awarded the Medal of Honor for his stout-hearted defense on 28 November 1950 of his position at Toktong Pass despite his repeated grievous wounds.

Captain William E. Barber

Captain William E. Barber, 31, of Dehart, Kentucky, commanding officer of Company F, 2d Battalion, 7th Marines, awarded the Medal of Honor for his intrepid defense of Toktong Pass from 28 November to 2 December in spite of his own severe wounds.

Private First Class William B. Baugh

Private First Class William B. Baugh, 20, born in McKinney, Kentucky, a member of the Anti-Tank Assault Platoon, Weapons Company, 3d Battalion, 1st Marines, posthumously awarded the Medal of Honor for covering with his body an enemy grenade thrown into the truck in which his squad was moving from Koto-ri to Hagaru-ri on the night of 29 November 1950 as part of Task Force Drysdale.

Major Reginald R. Myers

Major Reginald R. Myers, 31, of Boise, Idaho, executive officer of the 3d Battalion, 1st Marines, awarded the Medal of Honor for leading a hastily organized provisional company of soldiers and Marines in the critical assault of East Hill at Hagaru-ri on 29 November 1950.

Captain Carl L. Sitter

Captain Carl L. Sitter, 28, born in Syracuse, Missouri, commanding officer of Company G, 3d Battalion, 1st Marines, awarded the Medal of Honor for his valiant leadership in bringing his company from Koto-ri to Hagaru-ri as part of Task Force Drysdale on 29 November. He then led it in the continued assault of vital East Hill on 30 November 1950.

Staff Sergeant William G. Windrich

Staff Sergeant William G. Windrich, 29, born in Chicago, Illinois, a platoon sergeant with Company I, 3d Battalion, 5th Marines, posthumously awarded the Medal of Honor for his extraordinary bravery in taking and then holding a critical position near Yudam-ni on 29 November despite two serious wounds which eventually caused his death.

Lieutenant Colonel Raymond G. Davis

Lieutenant Colonel Raymond G. Davis, 35, of Atlanta, Georgia, commanding officer of the 1st Battalion, 7th Marines, awarded the Medal of Honor for his conspicuous gallantry and skill in leading his battalion, from 1 to 4 December 1950, across mountainous and frigid terrain to come to the relief of the beleaguered company holding Toktong Pass.

Sergeant James E. Johnson

Sergeant James E. Johnson, 24, of Washington, D.C. and Pocatello, Idaho, regularly a member of the 11th Marines but serving as a platoon sergeant of a provisional rifle platoon attached to the 3d Battalion, 7th Marines, posthumously awarded the Medal of Honor for continuing to engage the enemy single-handedly in hand-to-hand combat on 2 December south of Yudam-ni after being severely wounded.

Photo by Sgt Frank C. Kerr, National Archives Photo (USMC) 127-N-A5754

LtCol Murray's 5th Marines found a tent camp waiting for them when they arrived at Hungnam after the long march. Much of the camp was the work of the 1st Amphibian Tractor Battalion. Although there was still snow on the ground, the weather at Hungnam was mild in comparison with the sub-zero temperatures of the Changjin plateau.

GHQ and pick up his duties as MacArthur's chief of staff or he could remain in command of X Corps. Almond replied that he wished to stay with X Corps even if it became part of Eighth Army. The 27th of December was set as the day that X Corps would pass to Eighth Army control.

The evacuation of the 1st Marine Division began with the loading-out of the 7th Marines in the MSTS *Daniel I. Sultan*. The 5th Marines would follow on the 12th and the 1st Marines on the 13th. It was anticipated that the ships would have to make a second, even a third turn-around, to lift the entire division. The docks could berth only seven ships at a time. To compensate, there would be some double berthing, but most of the passengers would have to load out in the stream. Approximately 1,400 vehicles had been brought down from the Chosin plateau, about the same number as had gone up, but now some of the complement bore U.S. Army markings. Most of the division's vehicles would go out in LSTs. Green Beaches One and Two could handle 11 LSTs simultaneously. Thankfully there was no great tide to contend with, only one foot as compared to Inchon's 30.

Marine Close Air Support at Its Finest

The Marines' ground control intercept squadron, MGCIS-1, Major Harold E. Allen commanding, shut down at Yonpo on the 11th, passing control of air defense of the perimeter to the *Mount McKinley*. The sky remained empty of enemy aircraft. Overall air control stayed with MTACS-2, the Marines' tactical air control squadron, under Major Christian C. Lee. Each of the infantry battalions had gone into the Chosin campaign with two forward air controllers assigned, most of them reserves, all of them qualified Marine aviators. They brought with them both expertise in close air support and rapport with the fighter-bombers overhead. Although inclined to lament not having a cockpit assignment they realized they were providing an unmatched link to the air. They knew how to talk a pilot onto a target.

Between 1 and 11 December, Marine aviators, ashore and afloat, flew more than 1,300 sorties in support of their comrades on the ground. Of these, 254 were flown from the *Badoeng Strait* and 122 from the late-arriving *Sicily*. (Lundin's Blacksheep squadron, still at Wonsan, reembarked in the

MajGen Almond gave his Marine deputy chief of staff, Col Edward H. Forney, much of the credit for the orderly departure of X Corps from Hungnam. Here, on 14 December, he presents Forney with a Legion of Merit.

National Archives Photo (USA) 111-SC355068

Photo by Sgt Jack T. McKirk, National Archives Photo (USA) 111-SC354464

MajGen Almond, left, always generous with medals and commendations, on 11 December congratulates bareheaded BGen Armistead D. Mead, commander of Task Force Dog, on keeping the MSR open from Chinhung-ni to Hamhung. Few Marines were aware of this Army contribution to their march back from Chosin.

of ice and snow from the flight deck. The *Sicily* at one point had to stop flight operations for VMF-214's Blacksheep in the face of heavy seas and 68-knot winds. Planes were lost. Three night fighters went down. There were other crashes. It was estimated that a pilot who had to ditch at sea in the arctic waters had only 20 minutes before fatal hypothermia. Two VMF-212 pilots from Yonpo, out of gas, managed to save themselves and their planes by landing on the *Badoeng Strait*. By strenuous effort on the part of all hands, aircraft availability at Yonpo hovered around 67 percent and a remarkable 90 percent on board the carriers. About half the missions flown were not for the Marines but for someone else. Statistics kept by the wing reported a total of 3,703 sorties in 1,053 missions controlled by tactical air control parties being flown between 26 October and 11 December. Of these missions, 599 were close support—468 for the 1st Marine Division, 67 for the

Sicily on 7 December.) The rest had been by the shore-based squadrons at Wonsan and Yonpo. The first Marine jet squadron to arrive in Korea, VMF-311, with McDonnell F9F Panther jets, Lieutenant Colonel Neil R. McIntyre commanding, had arrived at Yonpo on 10 December and managed to fly four days of interdiction missions before moving back to Pusan to aid the Fifth Air Force in its support of the Eighth Army.

Flight conditions both ashore at Yonpo and afloat in the carriers were hellish—in the air, poor charts, minimal navigational aids, and capricious radios; at Yonpo, primitive conditions and icy runways; and, afloat, ice-glazed decks and tumultuous seas for the carrier-based aircraft. The *Badoeng Strait* reported scraping off three inches

Gen MacArthur made one of his quick trips to Korea on 11 December, this time to Yonpo Airfield to meet with MajGen Almond and approve the X Corps evacuation plan. MacArthur is in his trademark peaked cap and Almond is in a bombardier's leather jacket. No one would mistake the accompanying staff officers, with their well-fed jowls, for combat soldiers.

National Archives Photo (USA) 111-SC3544110

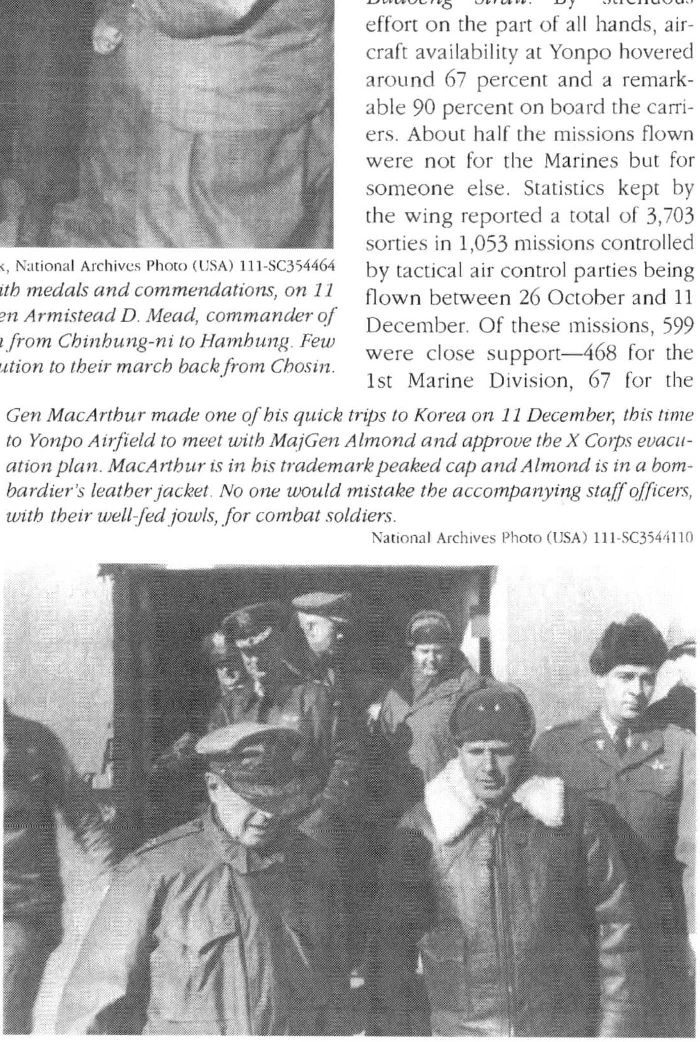

Photo by SSgt Ed Barnum, Department of Defense Photo (USMC) A130426

Marine crewmen check the loading of rockets on the rails of an F-4U Corsair flying from Yonpo Airfield early in December. Despite minimum facilities and terrible weather, the hard-working crews maintained an availability rate of 67 percent for aircraft flying from Yonpo.

ROKs, 56 for the 7th Infantry Division, and 8 for the 3d Infantry Division. Eight Marine pilots were killed or died of wounds, three were wounded, and four were missing in action.

Marine transports—twin-engine R4Ds and four-engine R5Ds from VMR-152, commanded by 44-year-old Colonel Deane C. Roberts—supplemented General Tunner's Combat Cargo Command in its aerial resupply and casualty evacuation from Hagaru-ri.

The squadron that the 1st Marine Division considered its own private air force, Major Gottschalk's VMO-6, with 10 light fixed-wing aircraft and nine helicopters, racked up 1,544 flights between 28 October and 15 December. Of these 457 had been reconnaissance, 220 casualty evacuation, and 11 search-and-rescue.

Time To Leave

Wonsan closed as a port on Sunday, 10 December. Outloading for the evacuation, conducted from 2 to 10 December, was under Lieutenant Colonel Henry "Jim" Crowe with muscle provided by his 1st Shore Party Battalion. The attached Company A, 1st Amphibian Truck Battalion, found employment for its DUKWs (amphibian trucks) in shuttling back and forth between docks and ships. In the nine-day period, 3,834 troops (mostly Army), 7,009 Korean civilians, 1,146 vehicles, and 10,013 tons of bulk cargo were evacuated. Defense of the immediate harbor area was shared with two battalions of South Korean Marines and a battalion from the 3d Infantry Division.

General Craig, Smith's sorely missed assistant division commander, returned from emergency leave on the 11th. Marines were left to wonder what his tactical role might have been if he had come back earlier. Smith sent him south to Pusan to arrange for the division's arrival. "I took 35 people of various categories with me and left for Masan," said Craig years later. "[I] conferred with the Army commander there about replacement of enormous losses of equipment of various kinds. He assured me that he would open his storerooms to us and give us anything we required that was in his stock. And this he did."

On Tuesday evening, 12 December, General Almond called his generals together for a conference and a dinner at X Corps headquarters. The division commanders—Smith, Barr, and Soule—listened without comment to a briefing on the evacuation plan. They then learned that the true purpose of the dinner was Almond's 58th birthday. General Ruffner, Almond's chief of staff, eulogized his commander, saying, in effect, that never in the history of

Nerve center for the tactical air support of X Corps was the 1st Marine Aircraft Wing's Tactical Air Control Center at Hambung. After the last of the Marine squadrons departed Yonpo on 14 December, control of air operations passed to Navy air controllers on the command ship Mount McKinley *(AGC 7).*

Photo by SSgt Ed Barnum, National Archives Photo (USMC) 127-N-A130453

National Archives Photo (USN) 80-G-422892

Wings folded, Marine F-4U Corsairs wait on the ice-glazed deck of the escort carrier Badoeng Strait *(CVE 116). As much as three inches of ice had to be scraped from the flight deck. A remarkable aircraft availability rate of 90 percent was maintained on board the carriers. Only half the missions were in support of the Marines. The rest went to the Army and South Koreans.*

the U.S. Army had a corps in such a short time done so much. General Almond replied and General Shepherd added a few complimentary remarks. Earlier Almond had asked Smith if he thought it feasible to disinter the dead buried at Hungnam. Smith did not think it feasible.

Interdiction fires by Army artillery, deep support by naval gunfire, and air interdiction bombing by Air Force, Navy, and Marine aircraft provided a thunderous background of noise for the loading operations. By 13 December the 5th and 7th Marines were loaded and ready to sail. At 1500, General Smith closed his division command post ashore and moved it to the *Bayfield* (APA 33). Before departing Hungnam, Smith paused at the cemetery to join a memorial service for the dead. Protestant, Catholic, and Jewish chaplains officiated. Volleys were fired and taps sounded. Meanwhile, the 3d and 7th Infantry Divisions had nothing to report except light probing of their lines and minor patrol actions.

The loading of the Marines and attached Army elements was completed on the 14th. That day saw the last of the Marine land-based fighter-bombers depart Yonpo for Japan. Shortly after midnight the air defense section of MTACS-2 passed control of all air to the Navy's Tactical Air Control Squadron One on board the *Mount McKinley*, but, just to be sure, a standby Marine tactical air control center was set up on an LST and maintained until the day before Christmas.

The *Bayfield*, an attack transport and the veteran of many landings, with General Smith embarked, lifted her hook and sailed at 1030 on 15 December. The ship had been experimenting with C-rations, but with the embarkation of the Marines she

In a landing exercise in reverse, Marines in an LCM landing craft head for a transport waiting for them in Hungnam harbor. The docks could only berth seven ships at a time, so most soldier and Marines had to load out in the stream. Collecting enough ships, both U.S. Navy and Merchant Marine, for the evacuation was a monumental effort.

National Archives Photo (USN) 80-G-424506

Department of Defense Photo (USA) SC355243

These members of the 5th Marines move by way of a cargo net from an LCM landing craft into a side hatch of the transport that would take them to the southern tip of the Korean peninsula. Loading out of the regiment was essentially accomplished in one day, 12 December. Destination was the "Bean Patch" at Masan.

returned to a more appetizing Navy diet. A total of 22,215 Marines had boarded an assemblage of 4 transports, 16 landing ships, an assault cargo ship, and 7 merchant ships. General Shepherd, with his Marines safely embarked, left Hungnam the same day for Hawaii by way of Tokyo. Just before leaving Hungnam he attended a ceremony at which General Almond presented a Distinguished Service Cross to General Barr.

A day's steaming on board the jam-packed ships took the Marines to Pusan. They landed at Pusan and motor-marched to the "bean patch" at Masan where a tent camp was being set up. Smith moved into a Japanese-style house. "The toilet works, but the radiators are not yet in operation," he noted.

The Commandant reported to the Secretary of the Navy 4,418 Marine casualties for the period 26 October to 15 December. Of these, 718 were killed or died of wounds, 3,508 wounded, and 192 missing in action. In addition, there were 7,313 non-battle casualties, mostly frostbite. Roughly speaking, these non-battle casualties added up to a third of the strength of the division. (From 26 November until 11 December, Commander Howard A. Johnson's 1st Medical Battalion had treated 7,350 casualties of all categories.) The three infantry regiments had absorbed the lion's share of the casualties and arrived at the Bean Patch at about 50 percent strength. Some rifle companies had as little as 25 or 30 percent of their authorized allowance.

Chinese Casualties

Captured documents and prisoner interrogations confirmed that the Marines had fought at least nine and possibly all 12 CCF divisions. These divisions can be assumed to have each entered combat at an effective strength of about 7,500—perhaps 90,000 men in all. Other estimates of Chinese strength go as high as 100,000 or more. Peng's chief of staff said, it will be remembered, that the *Ninth Army Group* had started across the Yalu with 150,000 troops, but not all of these had come against the 1st Marine Division. The Marines could only guess at the casualties they had inflicted. The estimates came in at 15,000 killed and 7,500 wounded by the ground forces and an additional 10,000 killed and 5,000 wounded by Marine air.

Still waiting in the surrounding hills above Hamhung, Sung Shilun's *Ninth Army Group*—assuming non-combat casualties at least equal to battle casualties—probably had at most no more than 35,000 combat effectives. Almond's X Corps had three times that number. Rank-and-file Marines who grumbled, "Why in the hell are we bugging out? Why don't we stay here until spring and then counterattack?" may have had it right.

Last Days of the Evacuation

The light carrier *Bataan* (CVL 29) joined Task Force 77 on 16 December, too late to help the

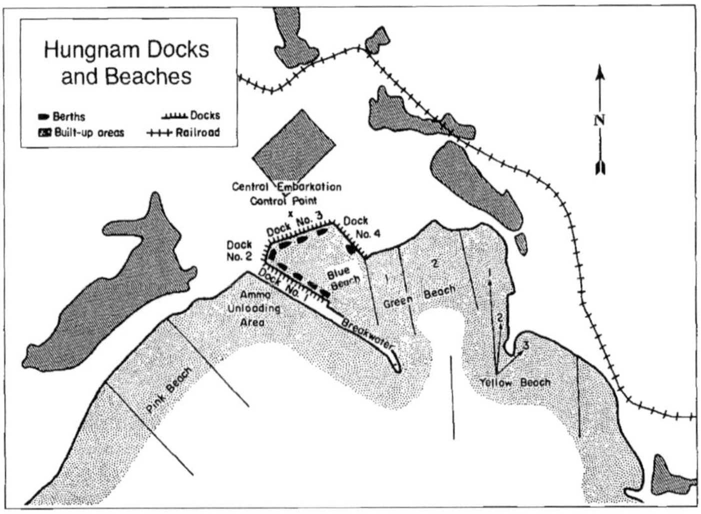

Photo by Cpl W. T. Wolfe, Department of Defense Photo (USMC) A5414

MajGen Smith, a deeply religious man, paid a last visit to the division's Hungnam cemetery before boarding the Bayfield *(APA 33) on 13 December. A memorial service with chaplains of three faiths, Protestant, Catholic, and Jewish, was held. Volleys were fired and taps sounded. During the course of the Chosin Reservoir campaign, 714 Marines were killed or died of wounds.*

Marines, but in time for the last stages of the Hungnam evacuation. Airlift from Yonpo continued until 17 December after which that field was closed and a temporary field, able to handle two-engine transports, opened in the harbor area. The only Marine units still ashore were an ANGLICO (Air Naval Gunfire Liaison Company) group, a reinforced shore party company, and one-and-a-half companies of the 1st Amphibian Tractor Battalion manning 88 amphibian tractors. These Marines had been left behind to assist in the outloading of the remainder of X Corps. General Smith had resisted this detachment, and General Shepherd, before departing, had advised Smith to stress to X Corps the irreplaceable character of the tractors. Admiral Doyle, as a safeguard, had earmarked several LSDs (landing ship, docks) to lift off the tractor companies and their vehicles.

The last of the ROK Army units sailed away on the 18th. General Almond closed his command post ashore on 19 December and joined Admiral Doyle in the *Mount McKinley*. Doyle reminded Almond that, in accordance with amphibious doctrine, all troops still ashore were now under his command as amphibious task force commander. By the 20th all of the 7th Infantry Division was embarked. On the morning of 24 December the 3d Infantry did its amphibious landing in reverse, coming off seven beaches into landing ships in smart style marred only by the premature explosion of an ammunition dump, set off by an Army captain, that killed a Marine lieutenant and a Navy sea-

Department of Defense Photo (USN) 424527

Marines, probably members of the division headquarters, board the Bayfield *by way of the gangway ladder. With MajGen Smith and his command group embarked, the* Bayfield *lifted her hook and sailed before noon on 15 December. A day's steaming would take the ship to Pusan.*

man and wounded 34 others. Three Marine amphibian tractors were lost in the explosion.

Totting up the statistics: 105,000 U.S. and ROK service men, 91,000 Korean refugees, 17,500 vehicles, and 350,000 measurement tons had gone out in 193 shiploads in 109 ships—some ships made two or even three trips.

The carrier *Valley Forge* came on station on 23 December, in time for the final curtain. By mid-afternoon on the 24th, all beaches were clear and the planned pyrotechnic display of demolitions and final naval gunfire began. The whole waterfront seemed to explode as prepared explosive charges went off, sending skyward such ammunition, POL, and other stores as could not be lifted off. On board the *Mount McKinley* the embarked brass enjoyed the show and then the command ship sailed away.

More naval shells were used at Hungnam than at Inchon. Navy records show that during the period 7 to 24 December the expenditure, headed off by 162 sixteen-inch rounds from the battleship *Missouri* (BB 63), included 2,932 eight-inch, 18,637 five-inch, and 71 three-inch shells plus 1,462 five-inch rockets. The Chinese did not choose to test seriously the Hamhung-Hungnam perimeter defenses. Not a man was lost to enemy action.

After the short run south, General Almond went ashore from the *Mount McKinley* at Ulsan at mid-afternoon with Admiral Doyle to inspect unloading areas. Late in the evening they returned in the admiral's barge to the flagship and then went ashore again for Christmas dinner, Doyle explaining to Almond that no alcoholic drinks could be served on board ship.

Chairman Mao Is Pleased

On 17 December the Chinese occupied Hamhung. On the 27th they moved into Hungnam. Chairman Mao sent the *Ninth Army Group* a citation: "You completed a great strategic task under extremely difficult conditions."

But the costs had been high. The assaults against Yudam-ni and Hagaru-ri had almost destroyed the *20th* and *27th CCF Armies*. From Koto-ri on most of the Chinese fight was taken up by the *26th CCF Army*.

Zhang Renchu, commander of the *26th CCF Army* lamented in his report:

> A shortage of transportation and escort personnel makes it impossible to accomplish the mission of supplying the troops. As a result, our soldiers frequently starve. From now on, the organization of our rear ser-

Department of Defense Photo (USA) SC355244

Vehicles of LtCol Youngdale's 11th Marines are swung up on board a merchant ship at a Hungnam dock on 14 December. Some ships had to make two or even three round trips before the evacuation was completed. About 1,400 vehicles had been brought down to Hungnam by the division. Most would go out by LSTs (tank landing ships).

vice units should be improved.

The troops were hungry. They ate cold food, and some had only a few potatoes in two days. They were unable to maintain the physical strength for combat; the wounded personnel could not be evacuated. . . . The fire power of our entire army was basically inadequate. When we used our guns there were no shells and sometimes the shells were duds.

Zhang Yixiang, commander of the *20th CCF Army*, equally bitter, recognized that communications limitations had caused a tactical rigidity:

Our signal communication was not up to standard. For example, it took more than two days to receive instructions from higher level units. Rapid changes of the enemy's situation and the slow motion of our signal communications caused us to lose our opportunities in combat and made the instructions of the high level units ineffective.

We succeeded in the separation and encirclement of the enemy, but we failed to annihilate the enemy one by one. For example, the failure to annihilate the enemy at Yudam-ni made it impossible to annihilate the enemy at Hagaru-ri.

Zhang Yixiang reported 100 deaths from tetanus due to poor medical care. Hundreds more were sick or dead from typhus or malnutrition to say nothing of losses from frostbite. The *26th CCF Army* reported 90 percent of the command suffering from frostbite.

An Army band greets Col Litzenberg's 7th Marines on arrival at Pusan. From here the regiment moved by motor march to Masan where an advance party had the beginnings of a tent camp ready for them. After a pause for Christmas, the rebuilding of the 1st Marine Division began in earnest.

National Archives Photo (USMC) 127-N-A5704

Photo by David Douglas Duncan

The 1st Marine Division was only a fraction of the total evacuation from Hungnam. In all, along with 105,000 U.S. and South Korean servicemen, 91,000 civilian refugees were evacuated. In materiel, 17,500 vehicles and 350,000 tons of all classes of equipment and supplies were taken out in 193 shiploads in 109 ships. "We never, never contemplated a Dunkirk," Admiral C. Turner Joy later said.

MajGen Robert H. "Shorty" Soule, USA, supervises the loading-out of the last elements of his 3d Infantry Division at Hungnam on 24 December. By then, MajGen Almond, the X Corps commander, considered the 3d Infantry Division, which had not been involved in heavy fighting, his most combat-effective division.
National Archives Photo (USA) 111-SC355587

Peng Deqing, commander of the *27th CCF Army*, reported 10,000 non-combat casualties in his four divisions:

> The troops did not have enough food. They did not have enough houses to live in. They could not stand the bitter cold, which was the reason for the excessive non-combat reduction in personnel. The weapons were not used effectively. When the fighters bivouacked in snow-covered ground during combat, their feet, socks, and hands were frozen together in one ice ball. They could not unscrew the caps on the hand grenades. The fuses would not ignite. The hands were not supple. The mortar tubes shrank on account of the cold; 70 percent of the shells failed to detonate. Skin from the hands was stuck on the shells and the mortar tubes.

In best Communist tradition of self-criticism, Peng Deqing deplored his heavy casualties as caused by tactical errors:

> We underestimated the enemy so we distributed the strength, and consequently the higher echelons were over-dispersed while the lower echelon units were over-concentrated. During one movement, the distance between the three leading divisions was very long, while the formations of the battalions, companies, and units of lower levels were too close, and the troops were unable to deploy. Further-

Department of Defense Photo (USN) 424093

North Korean refugees wait apprehensively to board U.S. Navy LST 845. Some 91,000 civilians were evacuated from Hungnam. This does not count the thousands of others who fled Hungnam and other North Korean ports in fishing boats and other coastal vessels. Family separations occurred that in future years would never be mended.

more, reconnaissance was not conducted strictly; we walked into the enemy fire net and suffered heavy casualties.

Zhang Renchu, commander of the *26th CCF Army* found reason to admire the fire support coordination of the Marines:

> The coordination between the enemy infantry, tanks, artillery, and airplanes is surprisingly close. Besides using heavy weapons for the depth, the enemy carries with him automatic light firearms which, coordinated with rockets, launchers, and recoilless guns are disposed at the front line. The characteristic of their employment is to stay quietly under cover and open fire suddenly when we come to between 70 and 100 meters from them, making it difficult for our troops to deploy and thus inflicting casualties upon us.

In a 17 December message to Peng Dehuai, Mao acknowledged that as many as 40,000 men had perished due to cold weather, lack of supplies, and the fierce fighting. "The Central Committee cherishes the memory of those lost." Peng asked for 60,000 replacements; it would be April before the *Ninth Army Group* again went into combat.

Christmas at Masan

At Masan on Christmas Eve, Olin Beall, the mustang commander of the 1st Motor Transport Battalion, wrote a letter to his old commanding officer, General Holland M. "Howlin' Mad" Smith, now retired and living in La Jolla, California:

An enormous stockpile of equipment and supplies, including rations, fuel, and ammunition had been built up at Hungnam. Much was evacuated, but even more would have to be destroyed. A detachment of Marines was left behind to help in the destruction. A Marine lieutenant was killed on 24 December in a premature explosion, probably the last Marine casualty of the campaign.

Department of Defense Photo (USA) SC 355021

Department of Defense Photo (USN) 424297

The Begor *(APD 127), a high-speed amphibious transport, lends its weight to the pyrotechnic display ashore on 24 December. In a final farewell to the stricken port there was a crescendo of planned demolitions and naval gunfire. RAdm Doyle and MajGen Almond watched the spectacle from the bridge of the* Mount McKinley.

I just thought that you might like to have a few words on first hand information from an ole friend and an ole timer. . . . I've seen some brave men along that road and in these hills, men with feet frozen, men with hands frozen still helping their buddies, men riding trucks with frozen feet but fighting from the trucks. . . .I think the fight of our 5th and 7th Regts, from Yudam-ni in to Hagaru-ri was a thing that will never be equaled. . . . Litzenberg [7th] and Murray [5th] showed real command ability and at no time did any of us doubt their judgment. The night we came out of Koto-ri the temperature was 27 below zero and still we fought. Men froze to their socks, blood froze in wounds almost instantaneously, ones fingers were numb inside heavy mittens. Still men took them off to give to a wounded buddy. . . . We are now in

Department of Defense Photo (USA) 426954

More naval shells were shot at Hungnam than at Inchon. Here the battleship Missouri *(BB 63) bangs away with its 16-inch guns. Altogether the Navy fired more than 162 sixteen-inch, 2,932 eight-inch, and 18,637 five-inch shells plus 1,462 five-inch rockets during the period 7 to 24 December.*

Masan in South Korea reoutfitting, training and getting some new equipment. I'm very, very proud to be able to say that in all our operation my Bn [1st Motor Transport Battalion] has lost only 27 trucks and every one of these was an actual battle casualty, so I think my boys did pretty good. . . . Oliver P. Smith and Craig make a fine team and we'd stand by them thru hell and high water.

An epidemic of flu and bronchitis swept through the tent camp at Masan. The Marines were treated with an early antibiotic, Aureomycin, in capsules to be swallowed the size of the first joint of a man's finger. The division rebuilt itself rapidly. Replacements—men and materiel—arrived. Some units found themselves with an "overage" of vehicles and weapons that had to be returned to the Army.

A refrigerator ship brought into Masan a planned double ration of Christmas turkey. Through some mix-up a second shipment of turkey and accessories arrived so that there were four days of holiday menu for the Marines. Working parties pretending to be patrols went up into the surrounding hills to cut pine trees to line the company streets of the tent camp. C-ration cans and crinkled tinfoil from cigarette packages made do for ornaments. Choirs were formed to sing Christmas carols. Various delegations of South Koreans, civilian and military, arrived at the camp with gifts and musical shows.

On Christmas Day, General Smith was pleased to note that attendance at church services was excellent. Afterward he held open house at his Japanese-style house for officers of sufficient rank—his special staff, general staff, and more senior unit commanders. First Lieutenant James B. Soper, serving at Sasebo, Japan, had sent the commanding general's mess a case of Old Grand-Dad bourbon. Mixed with powered milk, sugar, and Korean eggs it made a passable eggnog.

The irrepressible LtCol Olin L. Beall in a photo taken at Camp Pendleton in May 1951. Beall's exploits as commanding officer of the 1st Motor Transport Battalion, which lost nothing in his own telling, delighted MajGen Smith, himself a reserved and rather humorless individual.

National Archives Photo (USMC) 127-N-A215229

National Archives Photo (USMC) 127-N-A5848

On Christmas Day, MajGen Smith held open house for his senior officers and immediate staff at his Japanese-style quarters in Masan. Not all can be readily identified, but easily recognizable on the General's right are Cols Puller, McAlister, and Bowser. On Smith's immediate left, with pipe, is BGen Craig. In the middle of the kneeling row is LtCol Beall.

Drysdale's 41 Commando also held an open house. The British embassy in Tokyo had sent over a supply of Scotch whisky and mincemeat pies. Most of the guests were officers of the 1st and 5th Marines.

On 27 December, for the benefit of his log, General Smith added up his division's losses since the Inchon landing on 15 September:

Killed in action	969
Died of wounds	163
Missing in action	199
Wounded in action	5,517
Total	6,848
Non-battle casualties	8,900
Prisoners of war taken	7,916

On the 28th of December the division was placed once again under the operational control of X Corps, still commanded by Almond who would soon be promoted to lieutenant general. X Corps was now part of the Eighth Army, which had a new commander. General Walker had been killed when this jeep collided with a South Korean weapons carrier north of Seoul on 23 December. Lieutenant General Matthew B. Ridgway, known to the Marines as a fighting paratrooper in World War II, took his place. General Smith met him for the first time at a conference at X Corps headquarters on 30 December. Ridgway told his listeners that he wanted less looking backward toward the MSR, saying that when parachutists landed their MSR was always cut. Smith, not sure if this was praise or criticism, was nevertheless cautiously impressed by the new commanding general.

By the first of the year the 1st Marine Division would be ready to return to combat. There would be new battles to be fought—and won.

Dressed for the occasion of MajGen Smith's Christmas party, LtCol Murray is wearing an Army winter trench coat and Col Puller a brand-new M1943 field jacket. Both wear the highly prized cuffed Army boots, but O. P. Smith, his inseparable pipe clutched in his left hand, is wearing, as always, regulation leggings with his high-top "boondockers."

Department of Defense Photo (USMC) A5850

What Happened to Them?

CHARLES L. "GUS" BANKS, commander of the 1st Service Battalion, received a Navy Cross for his actions at Hagaru-ri. He retired in 1959 with a promotion to brigadier general in recognition of his combat decorations and died in 1988.

BOEKER C. BATTERTON, commanding officer of MAG-12, retired in 1958 with a promotion to brigadier general in recognition of his combat decorations. He died in 1987.

OLIN L. BEALL, commanding officer of the 1st Motor Transport Battalion, retired as a colonel, with both a Navy Cross and a Distinguished Service Cross for his actions at Chosin Reservoir. He died in 1977.

ALPHA L. BOWSER, JR., the division's G-3 or operations officer, retired in 1967 as a lieutenant general and presently lives in Hawaii.

JAMES H. BROWER, commander of the 11th Marines, the artillery regiment, retired as a colonel in 1960 and died in 1984.

J. FRANK COLE, commanding officer of VMF-312, retired as a colonel in 1965 and died in 1969.

HENRY P. "JIM" CROWE, commanding officer of 1st Shore Battalion, retired in 1960 as a colonel, became chief of police in Portsmouth, Virginia, and died in 1991.

RAYMOND G. DAVIS, commander of the 1st Battalion, 7th Marines, went on to command the 3d Marine Division in Vietnam and was a four-star general and Assistant Commandant of the Marine Corps before retiring in 1972. He now lives near Atlanta, Georgia.

FREDERICK R. DOWSETT, the executive officer of the 7th Marines, retired as a colonel and died in 1986.

VINCENT J. GOTTSCHALK, commanding officer of VMO-6, received a Silver Star for his service in Korea. He retired as a colonel in 1968 and died in 2000.

FIELD HARRIS, commanding general of the 1st Marine Aircraft Wing, retired in 1953 and was advanced to lieutenant general because of his combat decorations. He died in 1967 at age 72.

WILLIAM F. HARRIS, commander of the 3d Battalion, 7th Marines, was listed as missing in action. No trace of him was ever found and he was eventually presumed dead. He received a posthumous Navy Cross.

BANKSON T. HOLCOMB, JR., the division's G-2 or intelligence officer, retired as a brigadier general in 1959. An expatriate, he lived for many years in Inverness, Scotland, where he died in 2000 at the age of 92.

MILTON A. HULL, company commander, Company D, 2d Battalion, 7th Marines, twice wounded, received both a Silver Star and Navy Cross for his actions. He retired as a colonel in 1969 and died in 1984.

ROBERT P. KELLER, commander of the "Blacksheep Squadron" and air liaison officer to Fifth Air Force, retired in 1974 as a lieutenant general. He lives in Pensacola, Florida.

RANDOLPH S. D. LOCKWOOD, commander of the 2d Battalion, 7th Marines, was not evacuated after being relieved but continued to move with the 7th Marines. On arrival at Masan he was sent to an Army hospital for psychiatric observation. The Army psychiatrist concluded he had suffered a situational neurosis, which disappeared after the evacuation. Lockwood returned briefly to the 2d Battalion, 7th Marines, but was soon transferred to administrative duties. He retired as a lieutenant colonel in 1960 and resides in Texas.

JAMES F. LAWRENCE, JR., who assumed command of 2d Battalion, 7th Marines, after Lockwood's relief, received a Navy Cross for his actions at the reservoir. After distinguished service as a Marine Corps lawyer, he retired in 1972 as a brigadier general. He lives in northern Virginia.

HOMER L. LITZENBERG, JR., commanding officer of the 7th Marines, rapidly ascended in grade to major general and as such in 1957 served as the senior member of the United Nations component negotiating the peace talks at Panmunjom. He retired in 1959, was elevated to lieutenant general because of his combat decorations, and died in 1963 at age 68.

FRANCIS M. MCALISTER, the division's G-4 or logistics officer, succeeded Puller as the commander, 1st Marines, a position he held until wounded in May 1951. He retired as a major general in 1960 and died in 1965.

JOHN N. MCLAUGHLIN, survived his captivity and went on to become a lieutenant general and chief of staff at Headquarters Marine Corps. He retired in 1977 and lives in Savannah, Georgia.

RAYMOND L. MURRAY, commander of the 5th Marines, rose to the grade of major general before retiring in 1968. He lives in Southern California.

Reginald R. Myers, Executive Officer, 3d Battalion, 1st Marines, received a Medal of Honor for his actions on East Hill. He retired as a colonel in 1967 and now lives in Florida.

George R. Newton, commander of the 1st Battalion, 5th Marines, received a Silver Star for his service in Korea and retired as a colonel in 1964. He died in 1993.

Francis F. "Fox" Parry, commander of 3d Battalion, 11th Marines, retired as a colonel in 1967. He published his memoir, *Three War Marine*, in 1987. He lives in Flourtown, Pennsylvania.

John H. Partridge, the division engineer, retired as a colonel in 1965 and died in 1987.

Lewis B. "Chesty" Puller, the 1st Marines' commanding officer, was promoted to brigadier general and became the division's assistant commander in February 1951. He received his fifth Navy Cross for his performance at the Chosin Reservoir and rose to the grade of major general on active service and to lieutenant general on the retired list when he retired in 1955. He died in 1971 at the age of 73.

J. Robert Reinburg, commander of VMF(N)-513, retired as a colonel in 1978 and died in 1997.

Thomas L. Ridge, Commanding Officer, 3d Battalion, 1st Marines, received a Silver Star for his defense of Hagaru. He retired as a colonel in 1964 and died in 1999.

Maurice E. Roach, Litzenberg's jack-of-all-trades, retired as a colonel in 1962 and died in 1988.

Deane C. Roberts, commander of VMR-152, retired as a colonel in 1957 and died in 1985.

Harold S. Roise, commander of 2d Battalion, 5th Marines, received two Navy Crosses for his heroic actions. He retired as a colonel in 1965 and died in 1991.

Webb D. "Buzz" Sawyer, Litzenberg's roving battalion commander, received two Silver Stars for his actions at Chosin Reservoir and a Navy Cross for later heroics during the Chinese spring counteroffensive in April 1951. He retired as a brigadier general in 1968 and died in 1995.

Henry W. "Pop" Seeley, Jr., retired as a colonel in 1963 with his last years of active duty as a highly regarded logistics officer. He lives in Florida.

Donald M. "Buck" Schmuck, Commanding Officer, 1st Battalion, 1st Marines, was later advanced to executive officer of the regiment. He retired in 1959 and because of his combat decorations was advanced in grade to brigadier general. He lives in Wyoming and Hawaii.

Carl L. Sitter, company commander, Company G, 3d Battalion, 1st Marines, recipient of a Medal of Honor, retired as colonel in 1970. He was a long-time resident of Richmond, Virginia, until his death in 2000.

Oliver P. Smith, Commanding General, 1st Marine Division, was promoted to lieutenant general in 1953 and given command of Fleet Marine Force, Atlantic. He retired in 1955 and for his many combat awards was raised in grade to four-star general. He died on Christmas Day, 1977, at his home in Los Altos Hills, California, at age 81.

Edward W. Snedeker, the division's deputy chief of staff, retired as a lieutenant general in 1963. In retirement he was known as a world-class stamp collector. He died in 1995.

Edward P. Stamford, the Marine tactical air controller with Task Force Faith, retired as a major in 1961 and lives in Southern California.

Allan Sutter, commander of the 2d Battalion, 1st Marines, retired as a colonel in 1964 and died in Orange, Virginia, in 1988.

Max J. Volcansek, Jr., commander of VMF(N)-542, retired in 1956 and was advanced in grade to brigadier general because of his combat decorations. He died in 1995.

Harvey S. Walseth, the division's G-1 or personnel officer, after recovering from his wounds, returned to the division to serve as deputy chief of staff and commanding officer, rear echelon. He retired in 1960 as a colonel and resides in Santa Barbara, California.

Erwin F. Wann, Jr., commander of the 1st Amphibian Tractor Battalion, retired as a colonel in 1965 and died in 1997.

Gregon A. Williams, the division's chief of staff, retired as a major general in 1954 and died in 1968.

David C. Wolfe, successor to Reinburg as Commanding Officer, VMF(N)-513, served as the head of the U.S. military mission in the Dominican Republic before retiring as a colonel in 1965. He died in 1992.

Carl A. Youngdale, who relieved Brower as Commanding Officer, 11th Marines, went on to command the 1st Marine Division in Vietnam and retired as a major general in 1972. He died in 1993.

About the Author

Edwin Howard Simmons, a retired Marine brigadier general, was, as a major, the commanding officer of Weapons Company, 3d Battalion, 1st Marines, throughout the Chosin Reservoir campaign. His active Marine Corps service spanned 30 years—1942 to 1972—during which, as he likes to boast, successively in World War II, Korea, and Vietnam he had command or acting command in combat of every size unit from platoon to division. A writer and historian all his adult life, he was the Director of Marine Corps History and Museums from 1972 until 1996 and is now the Director Emeritus.

Born in 1921 in Billingsport, New Jersey, the site of a Revolutionary War battle, he received his commission in the Marine Corps in 1942 through the Army ROTC at Lehigh University. He also holds a master's degree from Ohio State University and is a graduate of the National War College. A one-time managing editor of the *Marine Corps Gazette* (1945-1949), he has been widely published, including more than 300 articles and essays. His most recent books are *The United States Marine: A History* (1998), *The Marines* (1998), and a Korean War novel, *Dog Company Six*. He is the author of an earlier pamphlet in this series, *Over the Seawall: U.S. Marines at Inchon*.

He is married, has four grown children, and lives with his wife, Frances, at their residence, "Dunmarchin," two miles up the Potomac from Mount Vernon.

THIS PAMPHLET HISTORY, one in a series devoted to U.S. Marines in the Korean War era, is published for the education and training of Marines by the History and Museums Division, Headquarters, U.S. Marine Corps, Washington, D.C., as part of the U.S. Department of Defense observance of the 50th anniversary of that war. Editorial costs have been defrayed in part by contributions from members of the Marine Corps Heritage Foundation. To plan and coordinate the Korean War commemorative events and activities of the Sea Services, the Navy, Marine Corps, and Coast Guard have formed the Sea Services Korean War Commemoration Committee, chaired by the Director, Navy Staff. For more information about the Sea Services' commemorative effort, please contact the Navy-Marine Corps Korean War Commemoration Coordinator at (202) 433-4223/3085, FAX 433-7265 (DSN288-7265), E-Mail: HonorAndRemember@hqmc.usmc.mil, Website: www.history.usmc.mil.

KOREAN WAR COMMEMORATIVE SERIES

DIRECTOR OF MARINE CORPS HISTORY AND MUSEUMS

Colonel John W. Ripley, USMC (Ret)

GENERAL EDITOR, KOREAN WAR COMMEMORATIVE SERIES

Charles R. Smith

EDITING AND DESIGN SECTION, HISTORY AND MUSEUMS DIVISION

W. Stephen Hill, *Visual Information Specialist*

Catherine A. Kerns, *Visual Information Specialist*

NAVY-MARINE CORPS KOREAN WAR COMMEMORATION

Lieutenant Colonel Ward E. Scott II, USMCR, *Coordinator*

U.S. Marine Corps Historical Center
1254 Charles Morris Street SE
Washington Navy Yard DC 20374-5040
2002

PCN 19000410000

Sources

The official history, *The Chosin Reservoir Campaign* by Lynn Montross and Capt Nicholas A. Canzona, volume three in the five-volume series *U.S. Marine Operations in Korea, 1950-1953*, provided a starting place for this account. However, in the near half-century since this volume was published in 1957, there has been a great deal of new scholarship as well as release of classified records, particularly with respect to Chinese forces. This pamphlet attempts to benefit from these later sources.

With respect to Chinese forces, *The Dragon Strikes* by Maj Patrick C. Roe has been especially useful as have various articles by both Chinese and Western scholars that have appeared in academic journals. *The Changjin Journal*, the electronic newsletter edited by Col George A. Rasula, USA (Ret), has provided thought-provoking detail on the role of U.S. Army forces, particularly RCT-31, at the reservoir. The as-yet uncompleted work on the Hungnam evacuation by Professor Donald Chisholm has yielded new insights on that critical culminating event.

Books, some new, some old, that have been most useful include—listed alphabetically and not necessarily by worth, which varies widely—Roy E. Appleman, *East of Chosin* and *South to the Naktong, North to the Yalu*; Clay Blair, *The Forgotten War*; Malcolm W. Cagle and Frank A. Manson, *The Sea War in Korea*; T. R. Fehrenbach, *This Kind of War*; Andrew Geer, *The New Breed*; D. M. Giangreco, *War in Korea, 1950-1953*; Richard P. Hallion, *The Naval Air War in Korea*; Max Hastings, *The Korean War*; Robert Leckie, *The March to Glory*; Douglas MacArthur, *Reminiscences*; Francis Fox Parry, *Three War Marine*; Russell Spurr, *Enter the Dragon*; Shelby L. Stanton, *America's Tenth Legion*; John Toland, *In Mortal Combat: Korea 1950-1953*; Rudy Tomedi, *No Bugles, No Drums*; and Harry Truman, *Memoirs*.

The official reports that proved most helpful were the Far East Command's *Command Report, December 1950*; the 1st Marine Division's *Historical Diary for November 1950*; the Commander, Task Force 90's *Hungnam Redeployment, 9-25 December 1950*; and the Headquarters, X Corps, *Special Report on Chosin Reservoir, 17 November to 10 December 1950*.

Oral histories, diaries, memoirs (published and unpublished), and personal correspondence were extremely useful, especially those papers originating with Generals Almond, Bowser, Craig, Litzenberg, Murray, Shepherd, and Smith.

Resort was made to scores of biographical and subject files held by the Reference Section of the Marine Corps Historical Center.

The author also unabashedly put to use his own recollections of events and recycled materials that he had first developed on Chosin Reservoir in various essays, articles, and lectures during the past half-century.

As is invariably the case, the author had the unstinting and enthusiastic support and cooperation of the staff at the Marine Corps Historical Center.

The text has benefited greatly from the critical reviews by the editorial ladder within the Marine Corps Historical Center—Mr. Charles R. "Rich" Smith, Mr. Charles D. Melson, LtCol Jon Hoffman—and externally by Col Joseph Alexander, Col Thomas G. Ferguson, USA (Ret), BGen James F. Lawrence, Col Allan R. Millett, Mr. J. Robert Moskin, Col George A. Rasula, USA (Ret), and Maj Patrick C. Roe. The author, of course, remains responsible for any defects remaining in the book.

A fully annotated draft manuscript is on deposit at the Marine Corps Historical Center. Virtually all of the reference materials published and unpublished, used can be found at the Marine Corps Historical Center in Washington, D.C., or at the Marine Corps Research Center at Quantico, Virginia.

Made in United States
Orlando, FL
04 November 2026

90204190R00074